GOVERNORS STATE UNIVERSITY LIBRARY

W9-BSR-485

3 1611 00093 4759

QP
514.2
.Y83
1971

Yudkin, Michael **63128**
    A guidebook to
    biochemistry.

UNIVERSITY LIBRARY
Governors State University
Park Forest South, Il. 60466

A GUIDEBOOK
TO BIOCHEMISTRY

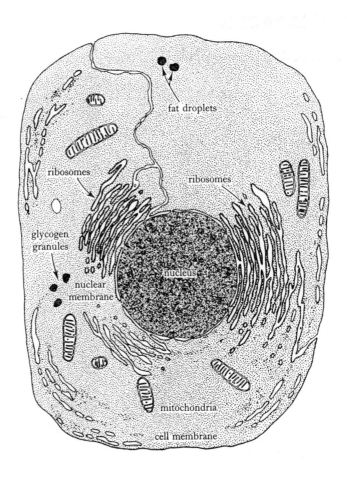

A typical animal cell, within which the majority of the events described in this book may take place. The function of the nucleus is discussed in Chs. 18, 19 and 21. The function of the mitochondrion is discussed in Ch. 8. The structure and function of the ribosome are described in Chs. 6 and 20. The synthesis of glycogen is described in Ch. 15 and that of fats in Ch. 16. The structure and function of glycogen are discussed in Ch. 6. The nature of the external cell membrane, the internal cell membrane (the endoplasmic reticulum) and the mitochondrial membrane is fundamentally uniform and is mentioned in Ch. 6.

A GUIDEBOOK TO BIOCHEMISTRY

by Michael Yudkin
Tutorial Fellow of University College, Oxford &
University Lecturer in Biochemistry

and Robin Offord
Fellow of University College, Oxford &
Senior Research Officer in Molecular Biophysics

a new edition of
A Guidebook to Biochemistry
by K. Harrison

CAMBRIDGE
AT THE UNIVERSITY PRESS 1971

University Library
GOVERNORS STATE UNIVERSITY

Published by the Syndics of the Cambridge University Press
Bentley House, 200 Euston Road, London NW1 2DB
American Branch: 32 East 57th Street, New York, N.Y.10022

Second edition *A Guidebook to Biochemistry*, K. Harrison
© Cambridge University Press 1965
This edition © M. Yudkin and R. Offord 1971

Library of Congress Catalogue Card Number: 70–153012

ISBN
0 521 08195 5 clothbound
0 521 09654 5 paperback

Printed in Great Britain
at the University Printing House, Cambridge
(Brooke Crutchley, University Printer)

QP
514.2
Y83
1971

Contents

Section III. Molecular genetics and protein synthesis

Conventions and abbreviations

Biochemical reactions commonly take place at or about pH 7. This poses the problem of how one should write the structural formulae of compounds which take part in acid-base equilibria. Such compounds will be wholly dissociated, partly dissociated or undissociated at pH 7 depending on the pK of their particular dissociation equilibrium.

To avoid confusion we have tried wherever possible to write the structures of the molecules we discuss in the un-ionized form, irrespective of their true pK and actual state of ionization at pH 7. Thus we write in equation 11, p. 93, of the synthesis of lactic acid when in fact what is produced is mainly lactate ion,

$$CH_3.CHOH.COO^-,$$

balanced by a solvated proton, H_3O^+. We depart from this rule in the few cases where it would hinder rather than help comprehension (e.g. p. 69).

In the case of the hydrogen carriers NAD and NADP we ignore ionization completely (see pp. 27 and 75) and write NAD and NADP for the oxidized form and $NADH_2$ and $NADPH_2$ for the reduced form.

These abbreviations are still allowed by international convention, although the symbols NAD^+, $NADP^+$, NADH and NADPH are now preferred. We have rejected this convention because we believe that it makes the events in, for example, Ch. 8 more difficult to follow.

We have followed the frequently adopted practice of neglecting the ionization of the phosphate group by writing (P) for phosphate in an organic compound and P_i for the inorganic phosphate ion.

Similarly, organic pyrophosphate is written $(\text{P}-\text{P})$ and the inorganic ion $(P-P)_i$. Thus the reaction fructose-1; 6-diphosphate   fructose-6-phosphate

(in which the possible ionizations of the phosphate groups are already neglected) is written on p. 127 as

Apart from the increased rapidity with which reactions can be written, this convention has the advantage that we need no longer write H_3PO_4 as a substrate or a product and give the impression that so many biochemical reactions use or generate a strong acid.

As a consequence of neglecting ionization equilibria, equations do not always balance as regards H and OH groups. Thus if we were not to neglect ionization the equation on p. 101 should be

$$CH_3.CO.COO^- + CO_2 + ATP = \underset{\underset{\displaystyle CH_2.COO^-}{|}}{CO.COO^-} + ADP + P_i$$

pyruvate ion   oxaloacetate ion

rather than

$$CH_3.CO.COOH + CO_2 + ATP = \underset{\underset{\displaystyle CH_2.COOH}{|}}{CO.COOH} + ADP + P_i$$

as written there. It will be seen that in the latter form there is a proton missing from the left-hand side. We do not put it in the equation in case we should give the impression that it takes part in the mechanism of the main reaction.

Reversible reactions are written $\rightleftharpoons$ in the usual way. However, in a few cases where the equilibrium constant is definitely known to be very much in favour of one direction rather than the other, this is indicated by ⇌ and ⇌ (see p. 67). It happens that both reactions used above as examples should have arrows of this sort and they are shown as such in their proper places in the text.

Where an overall reaction scheme involves *coupled* reactions (see p. 66) we have sometimes found it convenient to use the Baldwin notation

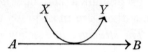

This notation is not meant to imply that the reaction scheme is mechanistically irreversible. It would be possible, but clumsy, to write the arrows for the return reaction.

Acknowledgments

Figs. 3.6 and 5.5.1 were specially drawn for this book by an ARGUS computer programmed and operated by Dr A. C. T. North, to whom the authors are most grateful. Figs. 4.4 and 6.1 were adapted from *The structure and action of proteins* by R. E. Dickerson and I. Geis (Harper and Row). The angle of perspective used in figs. 3.4 and 4.1.2 have been taken from those worked out by these authors for figures in their book, as have the representations of the amino-acid side chains in Table 3.1 (although the arrangement of the table is our own). The frontispiece is adapted from J. Brachet, *Sci. Amer.* **205**, 3 (1961). Fig. 3.3 is from *The nature of the chemical bond* by L. Pauling (Cornell Press). Fig. 6.2 is adapted from *The biochemistry of the nucleic acids* by J. N. Davidson (Methuen) and from M. F. H. Wilkins *et al.*, *Nature*, **175**, 834 (1955). Fig. 6.5 is an adaptation by J. N. Davison (*loc. cit.*) of a drawing by D. L. D. Caspar. Fig. 6.8 is from *Introduction to Lipids* by D. Chapman (McGraw-Hill). We thank authors and publishers for the use of this material.

1 Introduction

The principal difficulty in writing a book about biochemistry is in deciding the order of the chapters. In a linear subject like mathematics, the study of more advanced concepts depends on an understanding of more elementary concepts; there is a sequence in which one progresses, and books on mathematics adhere to the sequence. Biochemistry, by contrast, is a circular subject, in which the study of any one aspect can illuminate many other aspects; it seems to us that the authors of a book on biochemistry ought to make every effort to minimize the problems that arise from this characteristic of the subject.

We have tried to overcome the difficulties in several ways. In the first place the very idea of a guidebook is that it is intended to introduce the reader to just a few outstanding features of the field. This book is not intended to be comprehensive. We have omitted many topics that a textbook (even an elementary textbook) would normally include – to cite just a few examples, we have omitted any detailed discussion of specialization among the organelles of the cell, any discussion of experimental methods, any account of enzyme kinetics and all mention of some important synthetic reactions, e.g. the synthesis of urea. Our intention has been to introduce, exemplify and discuss certain crucial biochemical concepts. We have selected the topics that are best suited to the treatment of these even at the expense of seeming arbitrary in our choice of material.

The second way in which we have tried to deal with the circular nature of the subject is by dividing the book into three sections and introducing the key ideas of each section at the beginning. The first section is about the macromolecular constituents of living matter; it is introduced by a chapter on the forces involved in maintaining the structure of macromolecules. The second section is about intermediary metabolism; it is introduced by a chapter on energy, a chapter on oxidation, and a brief outline of the flow of carbon in

[1]

intermediary metabolism. The third section is about the synthesis of the informational macromolecules – DNA, RNA and protein – and the control of metabolic processes; it is introduced by a chapter on the genetic material (DNA) and its function in determining how proteins are synthesized.

Thus we begin and end the book with discussions of macromolecules, which are the characteristic compounds of living organisms; and this is an example of the circularity of the subject with which we are dealing. A further means that we have used in projecting this circular subject on to a book is to include extensive cross-references; we refer not only back but also forwards. And our final attempt to help the reader to break into the subject is connected with these cross-references; it is to express the hope that, as soon as you have finished the book, you will go back to the beginning and read it again.

Section I

STRUCTURE AND FUNCTION OF MACROMOLECULES

2 Introduction to macromolecules

Living matter is distinguished by its reliance on the special properties of certain classes of extremely large molecule. Of these classes the proteins, the nucleic acids and the polysaccharides are particularly prominent. All of these have common principles of construction, although these are at first sight obscured by the differences which exist as a necessary consequence of the great diversity of the functions they undertake.

The principal common feature is that all three are chain polymers formed by condensation, that is the combination of smaller molecules with the exclusion of water. In each case the smaller molecules are drawn from a homologous series. Proteins are composed of amino acids, of which the general formula is

$$R.CHNH_2.COOH$$

(see Ch. 3), polymerized by condensation between their amino and carboxyl groups. The resulting bond between the amino acids (which are now called amino acid residues) is known as the peptide bond. The end of the chain bearing the free amino group is called the amino terminus, that bearing the carboxyl group is called the carboxyl terminus.

Similarly nucleic acids consist of nucleotides. These are of general formula (purine or pyrimidine base) – pentose – phosphate (see Ch. 6) and they are joined by condensation between the phosphate group and an —OH group on the pentose of an adjacent nucleotide. The linkage between the nucleotide residues is called a phosphodiester bond. One end of the chain has a pentose in which the $3'$ position takes no part in the bonding and the other has a pentose in which the $5'$ position takes no part. These ends are called the $3'$ and $5'$ ends respectively (see p. 51).

Polysaccharides consist of sugars (p. 55) condensed through their —OH groups. The resulting bond between sugar residues is called the glycosidic linkage. It is an essential feature of the

[5]

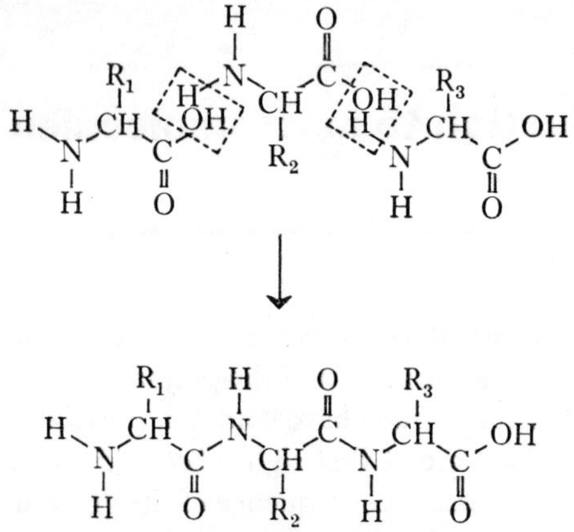

Fig. 2.1. The peptide bond.

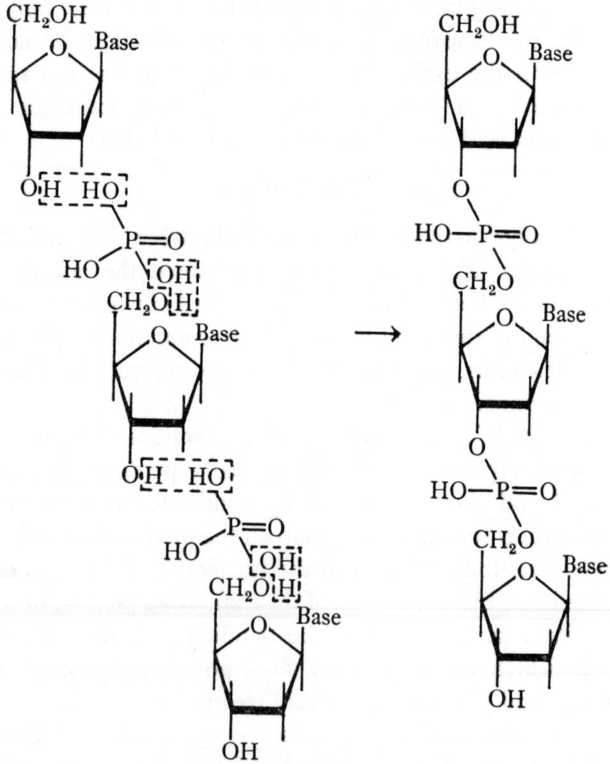

Fig. 2.2. The phospho-diester bond. The 5′ end of the chain
is at the top of the Figure.

glycosidic linkage that the C-1—OH of one of the sugars is involved. The other —OH group may belong to any of the carbon atoms of the second sugar. Because of the need for a C-1—OH in every bond only one end of the chain will have a sugar with position 1 free. The other end of the chain will always have a sugar with position 1 combined. Because of the reducing properties of the uncombined 1 position these ends are called the reducing and non-reducing ends respectively.

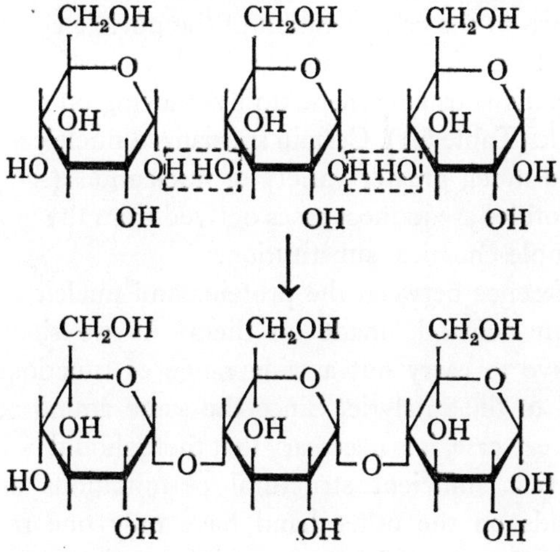

Fig. 2.3. The glycosidic bond. The reducing end is at the right.

It will be noticed that in each of these examples an unbranched polymer has been formed. However opportunities for cross-linking exist in each case. In the proteins the amino acids may possess groups in the side chains which can undergo cross-linking reactions. In fact this occurs only in a few of the possible cases (see p. 20), and in normal proteins never by formation of a peptide type bond. At points at which it does occur there can clearly be branching or joining of chains. In the nucleic acids the pentose may possess more than the two —OH groups shown and the theoretical possibility exists of branched or joined chains. They seem never to occur, however. Polysaccharides on the other hand are frequently branched, particularly when the molecule has as its main function the storage of sugars as food reserves. This sort of function is not critically dependent on structure and so the variation in structure from molecule to molecule caused by more or less random cross-linking of the chains can be tolerated. Bonds other than the glycosidic link are occasionally used for cross-linking (see p. 55).

It is when one looks at the nature of the homologous series involved, and the way in which selections are made from them to build up the molecule, that differences start to appear. The proteins have at their disposal about twenty different amino acids (Table 3.1) with a wide range of types of side chain. The difference when we turn to the consideration of the nucleic acids is striking. Here, once the choice of the pentose has been made (Table 6.1) the entire molecule is usually made up by drawing on only four types of nucleotide (Table 6.1). Certain specialized nucleic acids do exist which use a much greater variety of nucleotides (see p. 168), but these nucleotides are in most cases derived from the more common ones by simple chemical substitution.

This difference between the proteins and nucleic acids reflects the different demands made on them. As we shall see below proteins have to carry out a wide range of functions, from the mechanical to the catalytic. Since the same amino acids are involved in every case, it is necessary that there should be a reasonable range, so that sufficient structural permutations are possible. Nucleic acids on the other hand have only one main type of function, the storage and transfer of information (see pp. 50, 149). Here the permutation of just four types of smaller unit suffices.

The position with the polysaccharides is much as one would expect. The types of function usually undertaken here, mainly acting as structural material and for food storage (p. 56), do not call for any great subtlety in combination of units. Relatively few sugars are involved in the formation of the common polysaccharides. The most striking feature is that in many cases a given polysaccharide will draw on only one or occasionally two types of sugar molecule.

We see therefore that all these types of macromolecule are constructed by building up from sets of smaller molecules. For any given protein, nucleic acid or polysaccharide, how close is the control over the order in which their constituent residues are incorporated? The answer is surprising and was indeed thought at one time to be unbelievable. It appears that in the proteins and nucleic acids there is next to complete control. Barring accidents, a protein containing many hundreds of amino acids or a nucleic acid containing many thousands of nucleotides will be turned out by the synthetic machinery of the cell time and time again, without alteration. One species of protein or nucleic acid is absolutely distinguished from any other by its amino acid or nucleotide sequence. This tight control is essential because the functions of these macromolecules are sharply dependent on structure and even a small change in properties can be fatal to the delicate balance of physical and chemical events in living material.

The question of sequence does not arise in those polysaccharides which consist of a single sugar. Where more than one is used there is sometimes a measure of control of the sequence, so that if there are two, say, they may alternate along the chain. In other molecules where the desired function is less critically dependent on structure there is less control.

Subsequent chapters will describe a little of what is known about the structure of these macromolecules, the way in which structure influences function, and the way in which the organism ensures that the correct structure is obtained during synthesis of the macromolecules. These chapters will seek to show that the remarkable properties which distinguish living matter from non-living are very largely a result of the properties of macromolecules. These derive in their turn from recognizable features of the structure. The development of life had to await the discovery of ways of ensuring that the structures with beneficial properties could be repeated and safeguarded against deleterious changes from generation to generation.

The analogies between the three types of macromolecule become all the clearer if one takes the lipids (p. 56) as a contrary example. These molecules are not individually as large as macromolecules, though they are larger on the whole than the residues of the three classes we have covered. They do form assemblies which are macromolecular in type but which are joined, not by exclusion of water to form a covalent bond, but by non-covalent forces (see below). This gives them an entirely different character and questions such as the specificity of sequence do not arise.

Non-covalent interactions

The covalent bonds involved in the formation of the macro-molecules have been mentioned and we must now consider the non-covalent ones. As will be seen in the appropriate chapters these are just as important to the structure and function of the finished product as the covalent bonds. In fact, a look at the role of these non-covalent forces gives some clue as to why biological macro-molecules became necessary in the first place.

Little variation is possible in the strength, direction or other properties of any given covalent bond. On the other hand the properties of non-covalent bonds depend far more on environment. The variety of biochemical reactions and the need for their control and integration would easily exhaust the versatility of available covalent chemical interactions, numerous though these are. When small molecules are moving at random in free, dilute solution, the non-covalent interactions, which do have the required extra versatility, cannot be maintained in a useful form. However, when the interacting elements are anchored to one another in macro-molecules the situation changes dramatically. Combinations of non-covalent forces can now be produced that are useful, powerful, and capable of precise and infinite variation.

These combinations of forces and the local areas of high concentration of reactants that they make possible (pp. 45, 47) are the real foundation of the differences between biological and non-biological chemistry and thus of the differences between the living and the non-living state.

Potentially the strongest of the non-covalent interactions is the ionic bond, formed between ionized groups of opposite sign. In macromolecules most such groups are exposed to water (p. 12, but see p. 49 for an exception) and the ionic bond, like all electrostatic interactions, is screened as a result of the high dielectric constant of water and much weakened. They are, therefore, not now thought to be nearly so important in biochemistry as they were.

The hydrogen bond is next in order of potential strength. A simplified but adequate explanation for it is that a number of groups containing hydrogen, notably —OH and —NH_2, have an unequal distribution of electrons in which the hydrogen atom has less of them than its share. Other groups which do not contain hydrogen, $>$C$=$O for example, also have unequal distributions. A weak electrostatic interaction may be formed of the type

$$\overset{\delta-\;\;\delta+}{-\text{O}-\text{H}}\qquad\overset{\delta-\;\;\delta+}{\text{O}=\text{C}\diagdown}$$

which if at least one of the groups has a hydrogen atom is called a hydrogen bond.

Although hydrogen bond-forming groups are common in macromolecules (which is doubtless no coincidence) the role of the bond, though of the greatest significance, has still been somewhat overestimated in the past, since it also is subject to screening by water. However, suitable groups do exist in regions of macromolecules from which water is excluded and here, though still weak, they are numerous enough to make a vital contribution (e.g. p. 52) when taken together.

The hydrogen bond is a rather special case of so-called dipole forces. Many other charge inequalities, both permanent and temporary, exist in molecules found in biochemistry. Individually, they are weak and are effective only at very short range. It might therefore be thought that they are unimportant. However, it would be wrong to dismiss them for two reasons. The first is that there are a large number of dipoles in macromolecules and the resultant force obtained by summing all the dipole interactions might well be considerable. The second reason is that mathematical analysis shows that in many cases such a resultant, even though it derives from short range forces, is itself a *long* range force. In fact the resultant of a large number of dipole forces falls off more slowly than do forces obeying the inverse square law. Thus, though perhaps weaker initially, the resultant may outreach more normal electrostatic forces.

The last type of force we shall consider is the hydrophobic interaction. As two oil drops coalesce when they touch in water, so adjacent hydrophobic structures (non-polar molecules, paraffins or the side chains of certain amino acids for example) find that the closer they are together, the more stable the arrangement is (Fig. 2.4). The underlying theory of this common-sense conclusion is related to the hydrogen-bonding properties of water and cannot be discussed here.

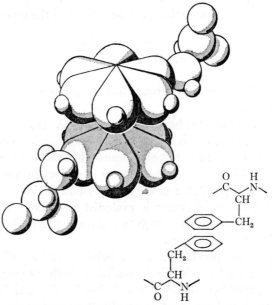

Fig. 2.4. The close approach of two hydrophobic side chains.

The hydrophobic bond is of prime importance to the structure and function of macromolecules (see Chs. 3, 4, 5 and 6). It plays a major part in maintaining the structure of both proteins (p. 22) and nucleic acids (p. 52). Both types of macromolecule have elements which favour hydrophobic bonding and elements which are hydrophilic. The hydrophobic elements will naturally lie, as far as is possible, in the centre of the structure, away from the solvent water. The hydrophilic elements will, on the contrary, lie on the surface of the molecule where they may interact with the water. The analogies with a detergent micelle (Fig. 2.5) are clear. Detergent molecules also have a hydrophobic region and a hydrophilic region. The former associate with each other and the latter interact with the water in precisely the same way as do the corresponding elements of macromolecules. This explains the overall similarity of Fig. 2.5 with the generalized picture of a protein (Fig. 2.6) and a nucleic acid (Fig. 6.3).

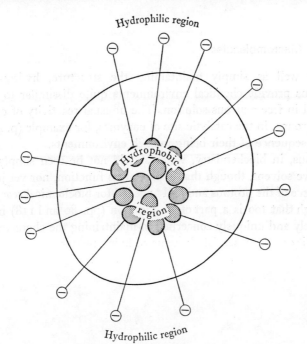

Fig. 2.5. A detergent micelle.

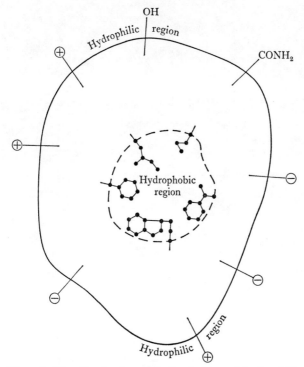

Fig. 2.6. The distribution of hydrophobic and hydrophilic regions in a protein molecule.

As well as simply maintaining the structure, hydrophobic regions provide chemical environments quite dissimilar to those found in free aqueous solution. The unusual reactivity of certain key groups, in the catalytic site of enzymes, for example (p. 49), is a consequence of their being in such environments.

Thus, in biochemistry, water should not be seen simply as a passive solvent, though that is part of its function, nor yet just as a source of the protons and OH^- needed for biochemical reactions, though that too is a part of its function (pp. 80 and 118) but as actively and uniquely concerned in maintaining the character and stability of the essential elements of living matter.

3 Protein Structure

As was stated in the last chapter the proteins differ from one another in the selection and arrangement of their constituent amino acids. The next two chapters show something of the types of function undertaken by proteins. This remarkable range is achieved either solely by permutation of amino acids or at most with the addition of one or two other small molecules which are not amino acids (see p. 28). We shall now examine the amino acids used in proteins from the point of view of the types of forces mentioned in the last chapter, and see what properties each brings with it to help in the task of producing a functioning molecule.

Proteins are of considerable size, usually having hundreds of amino acids, and they might therefore be expected to be sprawling ill-defined structures. In fact the majority of proteins are compact, highly convoluted molecules with the position of each atom relative to the others determined with great precision. An error in position of a constituent part of as little as the diameter of one atom may be sufficient to inactivate a protein. Thus we have to consider more than just the few residues which may contribute directly to the function of the protein by mediating in the interaction between the protein and other molecules (see next chapter). Many other residues will make as vital a contribution indirectly by maintaining the required precision in the structure of the protein itself. Both types of contribution involve the covalent and non-covalent interactions mentioned in the last chapter, and a major part of this chapter will be devoted to examining the way in which these interactions are employed in the stabilization of the structure. Their application to the external behaviour of proteins is best left to the chapter on protein function which follows.

Table 3.1 shows those amino acids that are normally found in proteins. There are many other naturally occurring amino acids but they do not appear in proteins. All, with the exception of proline, have their amino and carboxyl groups joined to the same carbon

Table 3.1. *The side chains (corresponding to R on p. 5) of the amino acids*

HYDROPHOBIC

No of carbon atoms in side chain

0	1	2	3	4	Cyclic

Glycine

Alanine

Valine

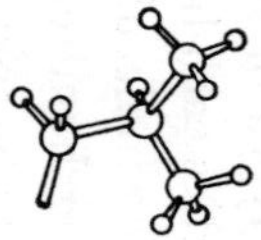

Leucine

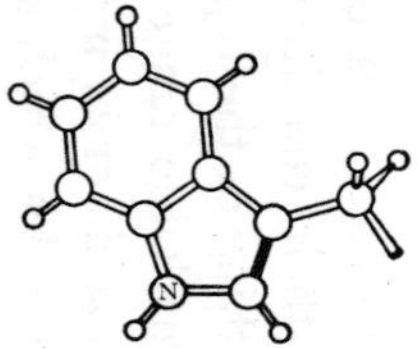

Isoleucine

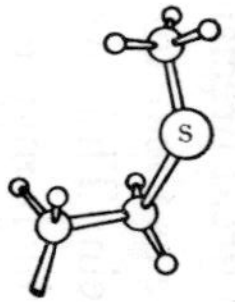

Methionine

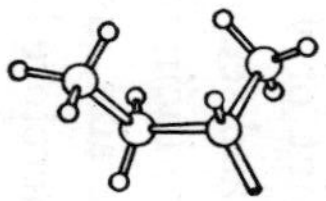

Tryptophan

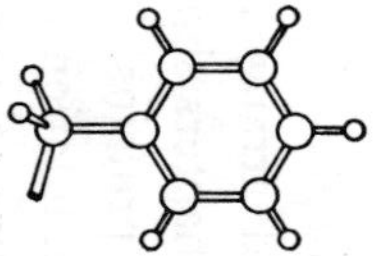

Phenylalanine

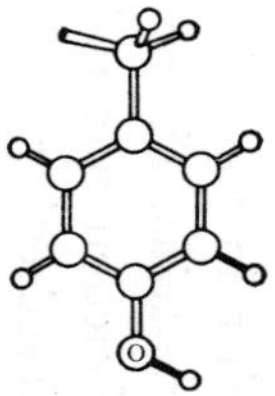

Tyrosine

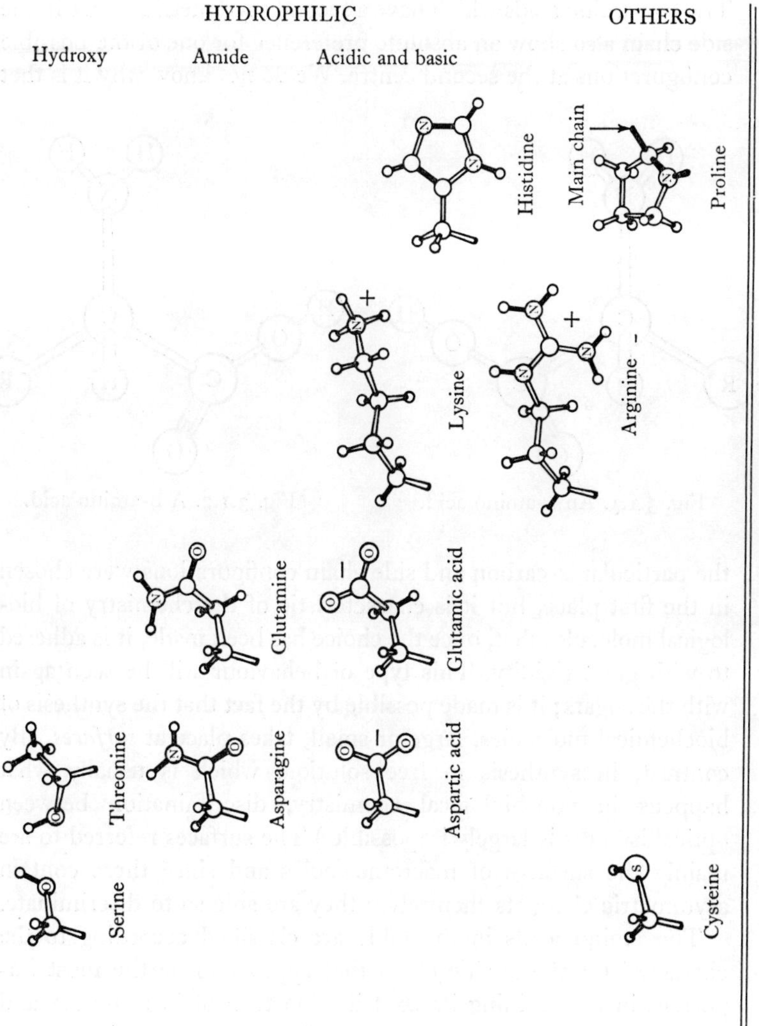

HYDROPHILIC OTHERS

Hydroxy   Amide   Acidic and basic

Histidine

Main chain

Proline

Lysine

Arginine

Glutamine

Glutamic acid

Serine   Threonine   Asparagine   Aspartic acid

Cysteine

atom (called the α-carbon). All have at least one centre of asymmetry, except glycine. The configuration about the asymmetric carbon atom is that shown in Fig. 3.1.1 as opposed to Fig. 3.1.2. The two amino acids which have a second asymmetric centre in the side chain also show an absolute preference for one of the possible configurations at the second centre. We do not know why it is that

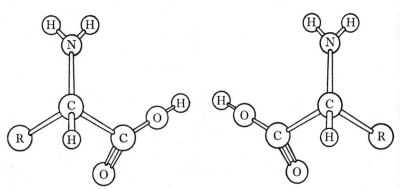

Fig. 3.1.1. An L-amino acid.   Fig. 3.1.2. A D-amino acid.

the particular α-carbon and side chain configurations were chosen in the first place, but it is characteristic of the chemistry of biological molecules that, once the choice has been made, it is adhered to with great rigidity. This type of behaviour will be seen again with the sugars; it is made possible by the fact that the synthesis of biochemical molecules, large or small, takes place at *surfaces*. (By contrast, in synthesis in free solution, which is usually what happens in non-biological chemistry, discrimination between optical isomers is largely impossible.) The surfaces referred to are mainly the surfaces of macromolecules and since these contain asymmetric elements themselves they are able so to discriminate.

The amino acids in the table are classified according to the characteristic of the side chain that appears to be the most important in determining its contribution to protein structure and activity. Members of the first group show a graduated increase of hydrophobic property as the size of the side chain increases. In the second group, those having hydrophilic side chains, there is again a graduated choice. Aspartic acid is more strongly acidic than is glutamic, histidine is a weaker base than lysine while arginine is a stronger one. Also, there is a spread of more generally hydrophilic, hydrogen-bonding properties. Cysteine (pronounced SIS-TAY-EEN) is in a class by itself in that it may be converted to cystine (SIS-TEEN) by oxidation (see Fig. 3.2). Proline is also in a class of its own since it is not truly an amino acid but an imino acid. The peptide bond formed by the $>$NH group of proline will clearly have geometrical and chemical properties which differ from those of the peptide bond formed by an NH_2 group. A consequence of this is, for example, that the α-helix (p. 21) cannot accommodate proline residues.

It must not be thought that the characteristics mentioned for each group are exclusive to that group or are its sole characteristics. Several amino acids not in the hydrocarbon group have some hydrophobic properties in parts of their side chain, e.g. the aliphatic hydrocarbon part of the lysine side chain (see p. 46). Another example is the ability of the —OH group of tyrosine and the —SH of cysteine to ionize. Several side chains are used in proteins for their chemical reactivity, cysteine and serine in particular.

The condensation of amino acids into the polypeptide chains by means of the peptide bond was discussed in the last chapter. The sequence of incorporation of amino acids, the so-called primary structure, is all important. Although a protein of 150 residues has about 10^{500} possible permutations, methods of analysis exist which make it possible to determine the true order in a remarkably short time.

We have already remarked that the properties of a protein depend on particular juxtapositions of side chains with the consequent interplay of the different types of characteristic mentioned. This juxtaposition is not merely one-dimensional, along the chain. The chain exists in three dimensions and it is its convolution to a precisely determined shape (tertiary structure) that allows the interplay of characteristics of side chains at its fullest. Proteins, which are therefore quite compact, are often divided into the categories 'globular' and 'fibrous'. An absolutely spherical shape is not required for a protein to be classed as globular, but when the ratio of length to width (known as the axial ratio) reaches or exceeds approximately 5:1 it is classified as fibrous.

Interactions stabilizing protein structure

The most important covalent interaction results from the oxidation of two cysteine residues to form cystine. This allows the joining or looping of polypeptide chains without the use of the peptide

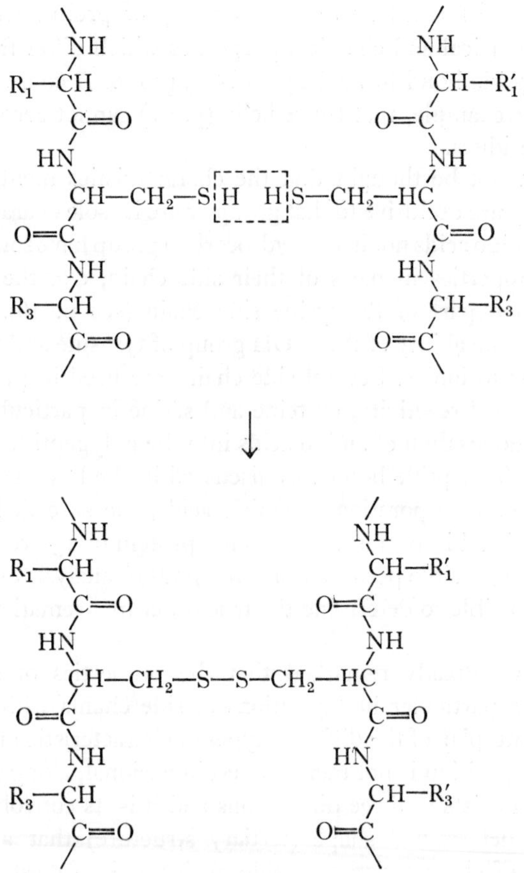

Fig. 3.2. Formation of the disulphide bridge.

bond (see Fig. 3.2). These so-called disulphide bridges are frequently found in proteins as a general aid to the stabilization of structure and they are also used where special mechanical properties are required (e.g. see p. 31). In exceptional circumstances cross-linking is brought about by other types of covalent bond, and these are mentioned on p. 31, too.

The ionic bond can occur by the interaction between on the one hand the positive charges on histidine, lysine, arginine and the α-amino group and on the other the negative charges of aspartic acid, glutamic acid and the α-carboxyl group. They are infrequently used in the stabilization of protein structure (see, however, Fig. 3.6), and are more frequently found in interactions between proteins and other molecules (see pp. 33 ff.).

The hydrogen bond is found between the side chains of the members of the third group of amino acids in Table 3.1. It is here that we note the peptide backbone itself is capable of a more active role in stabilizing the structure than simply holding the amino acids together. It has CO— and —NH groups regularly disposed along its length and these groups are eminently capable of hydrogen bonding to each other. As the groups are regularly arranged, it is not surprising that hydrogen bonding between them can give rise to regular structures. Features of this type are described as secondary structure and two, the α-helix and β-pleated sheet (Figs. 3.3 and 3.4) are found in most proteins. The reason why proteins are not entirely composed of elements with regular secondary structure is that there is a disruptive effect of other, over-riding interactions between certain of the side chains. However, the α-helix and the β-pleated sheet are quite common and are used as reinforcing members (struts and plates) in many proteins (Fig. 3.5). In a few cases they form a predominant part of the structure of a protein (see p. 30). The properties of the proteins will then depend to a significant extent on the properties of the regions of ordered secondary structure and these in their turn depend on the properties of the hydrogen bond. Hydrogen bonds are also of great importance in the interaction of proteins with other molecules (see pp. 35 and 48).

Other dipole forces must certainly be present in proteins since so many elements of the structure produce dipoles. They are thought to be quite important, especially in stabilizing quaternary structure (p. 25). Because of the large number of relatively feeble interactions which are involved little is known in detail of their contribution.

2

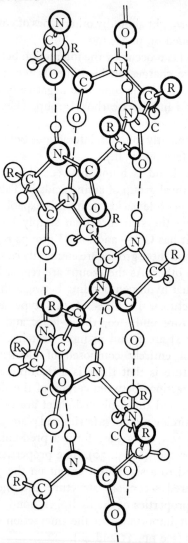

Fig. 3.3. The α-helix.

The hydrophobic interactions are probably the most powerful in stabilizing the structure and are also prominent among the forces involved in protein interaction with other molecules (see p. 35). Now that X-ray crystallography is making it possible to look at the tertiary structure of proteins it can be seen just how important these

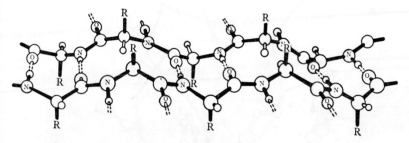

Fig. 3.4.   The β-pleated sheet.

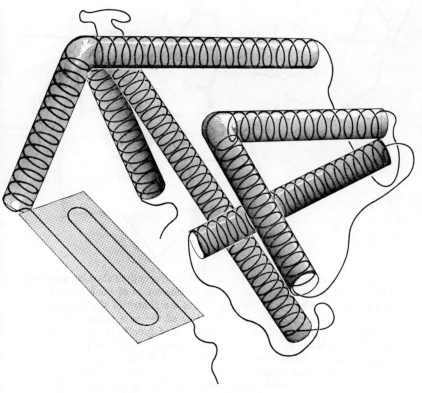

Fig. 3.5. Helical and sheet-like regions in a globular protein.

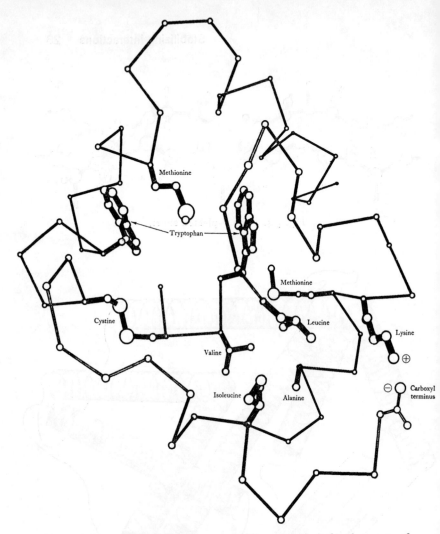

Fig. 3.6. A computer-drawn diagram (with perspective) showing some of the types of interaction stabilizing the conformation of the enzyme lysozyme (see also Fig. 5.5.1). Selected parts are shown of the polypeptide backbone of the protein and a very few of the side chains. Note the disulphide bridge (the cystine residue: one of four in the protein); the ionic bond between the NH_3^+ of the lysine and the COO^- of the carboxyl terminus of the protein (the only ionic bond in the protein); the close approaches between hydrophobic residues (just a few of those occurring in the complete structure). The clearest of the hydrophobic interactions is probably the 'sandwich' of a methionine side chain between two tryptophans. The atoms are not drawn to their full size for clarity; if they were it would be seen that many of the side chains are virtually touching.

are (Fig. 3.6). As suggested on p. 12 the molecule resembles a detergent micelle. The majority of the hydrophobic elements cluster together at the centre while only a few of them are exposed to the aqueous solution. The hydrophilic elements, on the other hand, are almost all exposed.

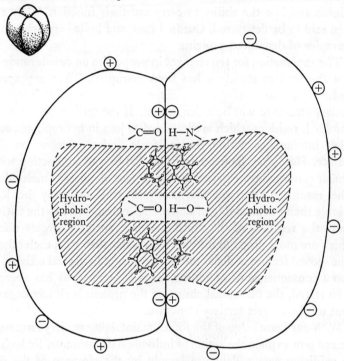

Fig. 3.7. Idealized section through a tetrameric arrangement of protein molecule (inset) showing types of interaction at an interface between two of the monomers.

Some proteins consist of aggregates of (usually) similar protein subunits. This so-called quaternary structure should not be thought of in terms of a random aggregate. The number of subunits involved and their geometry is precisely controlled. Some proteins rely on the extra possibilities conferred by a quaternary structure for the full expression of their activity (see pp. 39, 55 and 183). Quaternary structure is stabilized once again by the full range of non-covalent forces (see Fig. 3.7).

Denaturation

Proteins, as they are found in living tissue, are usually freely soluble materials with some definite chemical or mechanical function. When harshly treated (see below) they become far less soluble and lose the ability to carry out their function. They are then said to be denatured. Curdled milk and boiled egg white are examples of denatured protein.

The explanation for the changes in properties on denaturation is that the tertiary structure has been disrupted. Thus any agent leading to the weakening of any of the interactions maintaining the tertiary structure will be a denaturant. If the tertiary structure is unfolded, residues which are intended to join in hydrophobic and other interactions within the molecule find themselves on the surface. Here they have an equal probability of interacting with similar residues now exposed on the surface of adjacent molecules. The resulting intermolecular network will obviously be less soluble than the assembly of native molecules, since, in the native state, the residues of the micelle surface are very largely those which are more prone to interact with water than with each other. The second feature of denaturation, the loss of biological activity, is also a consequence of unfolding. This is because, as has already been stated, the functional ability of the protein is closely dependent on the correct tertiary structure.

With our knowledge of the forces maintaining tertiary structure we can now explain the action of individual denaturants. Reducing or oxidizing agents denature largely by the cleavage of the di-sulphide bridge; extremes of pH denature by charging or discharging ionizable groups; agents that interfere with hydrogen bonding will denature; detergents denature by disrupting the hydrophobic interactions; and so on. An increase in temperature leads, by increasing the random thermal motion of the constituents, to an increased likelihood of all bonds being broken.

Co-factors and prosthetic groups

The last feature to consider in this chapter is the class of small (non amino acid) molecules and ions which are occasionally found associated with the protein in its functioning state. They are usually called co-factors if weakly bound and prosthetic groups if strongly bound. (Certain substances may be co-factors or prosthetic groups depending on the particular protein. There is in any case no rigid line of distinction between the two terms.)

There are clearly tasks which are beyond even the quite respectable range of abilities of amino acid side chains, and it is essential for other molecules to be used on occasion. Co-factors and prosthetic groups, which are bound in sites specifically intended to enhance their usefulness, may behave in a way which would not be expected from their properties in free solution. This is analogous to the situation existing for the amino acid side chains themselves. The protein therefore accommodates the co-factors in two ways. It provides a specific binding site and it controls the chemical environment of that site.

Table 3.2 lists the more common co-factors and prosthetic groups. It will be seen that some of the structures given are quite complex. Many organisms, man included, are not able to synthesize all of these structures for themselves, and they must then be obtained in the diet. Many vitamins and so-called trace elements fall into this category. The very small dietary requirements for these substances can be explained by the fact that the proteins with the requirement for the co-factor or prosthetic group are very likely to be present themselves in only catalytic amounts. We can contrast this situation with the dietary requirement for the amino acids themselves. In man, for example, eight of the protein amino acids cannot be synthesized and have to be obtained in the diet. There is no question here of a need for trace amounts at a specific site, instead there is a general requirement for substances that are constituents of all proteins, and considerably larger amounts must be ingested.

Table 3.2. Co-factors and prosthetic groups

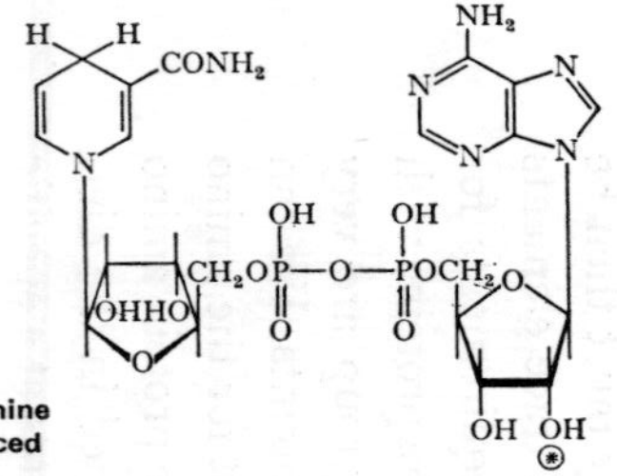

Nicotinamide Adenine Dinucleotide-reduced (NADH₂)
(see pp. 72–6)

NADP, NADPH₂ have ℗ at ⊛

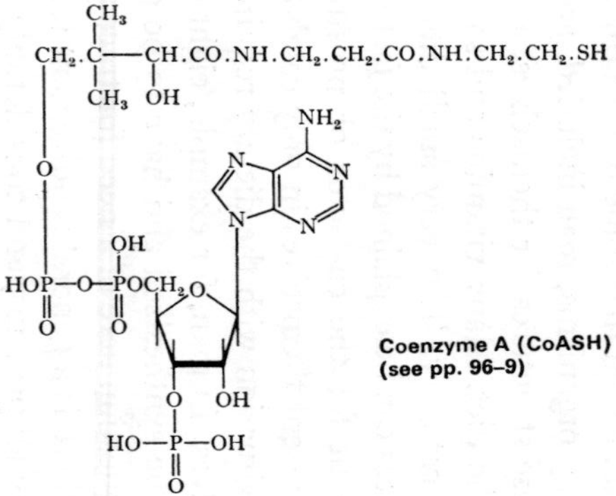

Coenzyme A (CoASH)
(see pp. 96–9)

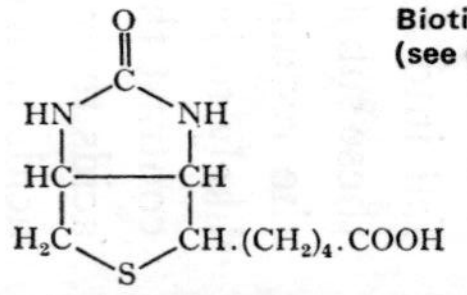

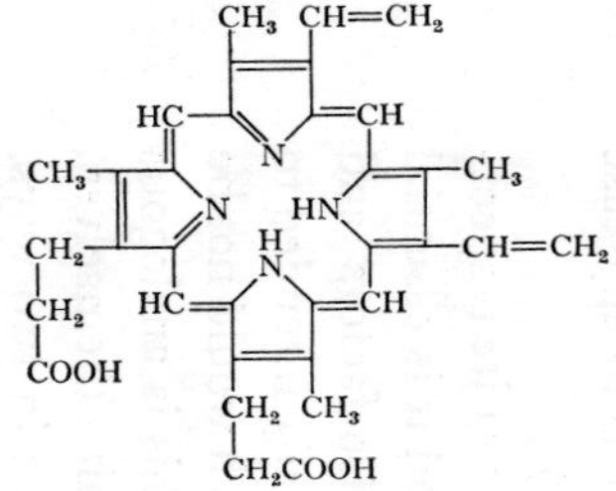

Porphyrin derivatives
(see pp. 37–9 and 76–7)

With Fe²⁺ : haemoglobin
Fe²⁺ ⇌ Fe³⁺ cytochromes
With Mg²⁺ chlorophyll
(Various side chains and substitutions)

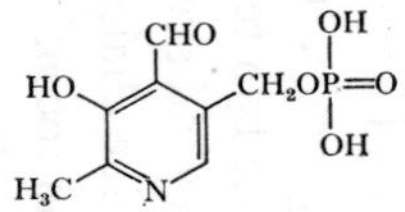

Pyridoxal phosphate
(see p. 137)

Riboflavin (vitamin B₂) and derivatives
(see p. 76)

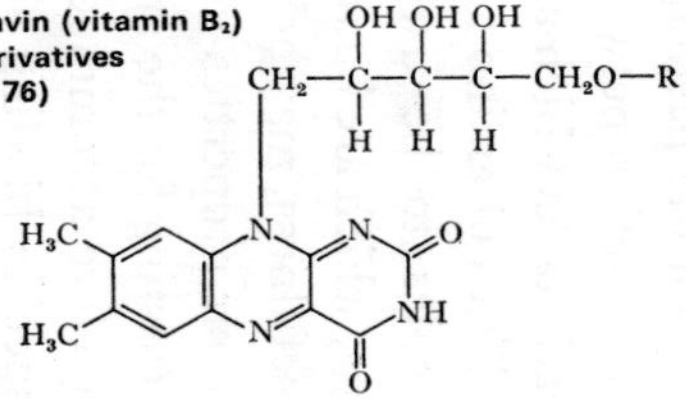

R=H: Riboflavin
R= ℗ Flavin mononucleotide (FMN)
R= —℗—℗—ribose-adenine Flavin adenine dinucleotide (FAD)

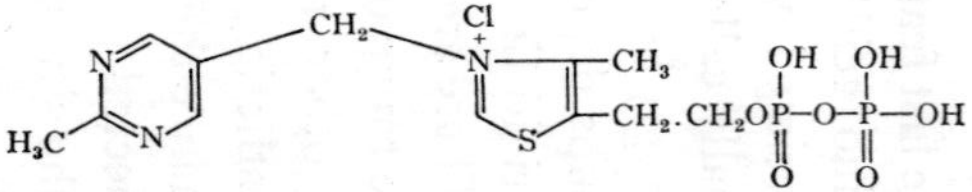

Thiamine pyrophosphate (TPP)
(see pp. 96–9)

Biotin (vitamin H)
(see p. 131)

4 Protein Function I

We will now see how it is that the same forces acting between the same set of elements manage to produce such a wide range of behaviour.

We may conveniently divide the proteins into three classes, depending on the way in which they contribute to the activities of living matter. The first class consists of the food storage proteins, the second of proteins which have a structural or mechanical role and the third of those which act by binding to other molecules.

Food storage proteins

These may be quickly dealt with. In order to synthesize a protein an organism must have an adequate supply of amino acids. In some cases it may be able to meet all its requirements by its own efforts, making use of the biosynthetic pathways described in Ch. 17. Should demand exceed this internal supply, or should an organism be totally incapable of synthesizing a particular amino acid (e.g. see p. 150), an external source is necessary. In such cases food proteins may be consumed, and then digested to amino acids by means of hydrolytic enzymes (p. 42). Almost any protein may serve as a source of amino acids in this way, but specialized molecules have been developed for particular purposes. For example, mammalian young are dependent on their mothers' milk for, among other things, a supply of amino acids to meet the severe nutritional problems posed by their early growth. This demand is met by a number of milk proteins, in particular by the caseins. Caseins, which are soluble until ingested, are found to be exceptionally easily denatured in the gut. This exposes many peptide bonds which are susceptible to enzyme-catalysed hydrolysis and the protein is readily digested. We shall see below (p. 40) that it is difficult to know when one has come to a full understanding of the function of a protein. However, as far as is known, this ease of denaturation and digestion is the major functional requirement for the casein molecule.

Structural and mechanical proteins

Elastic and inelastic structures

We consider now those proteins which act as more or less inert structural elements. These proteins are frequently of the high axial ratio (fibrous) type. They may be required to be elastic (for example, skin or hair) or relatively inelastic (for example, silk). α-Keratin is a good example of an elastic structure. It relies for both its strength and its elasticity on the α-helix, one of the regular arrangements of peptide backbone hydrogen bonds (Fig. 3.3) mentioned in the last chapter. It will be seen that the bonds, although weak individually, maintain a spring-like structure in which they are parallel and thus may be expected to have some collective strength (Fig. 4.1.1).

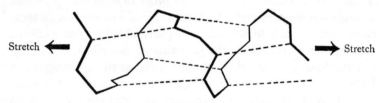

Fig. 4.1.1. Stretching an α-helix.

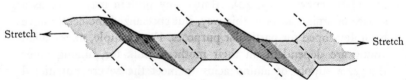

Fig. 4.1.2. Attempting to stretch a β-pleated sheet.

The inelastic protein α-fibroin occurs in certain silks. It typifies the β-pleated sheet structure (Fig. 3.4), another of the backbone hydrogen-bonded structures. Here the hydrogen bonds are at right angles to the direction of stretching and simply hold the bundle of adjacent polypeptide chains together. Extension is resisted by the full strength of the covalent bonds which in this case lie not perpendicular to but along the direction of stretching (Fig. 4.1.2). Collagen is another relatively inelastic fibrous protein. It has a remarkably high tensile strength and, weight for weight, is about as strong as steel. It has a complex structure based on a helical pattern.

Elasticity or rigidity may be enhanced by covalent cross-linking of the protein chain. A relatively open network of cross-links confers elasticity while a tighter, more numerous, system of cross-links confers rigidity (Fig. 4.2). The covalent bonds used may be S–S bridges (p. 20) or may arise in other ways, especially by means of special, aromatic free-radical reactions initiated by enzymes.

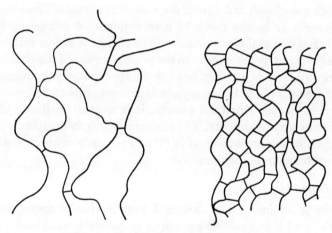

Fig. 4.2. Cross-linking of peptide chains.

In hair, the natural elasticity of α-keratin is enhanced by imbedding it in a matrix of S–S cross-linked protein chains. In horn, another keratin-containing structure, the matrix has more S–S bridges and is rigid. Certain elastic tendons, which rival in their properties the most advanced synthetic rubbers, do not have a matrix but use the aromatic free-radical method for direct cross-linking of the chain.

These free-radical reactions are also used in the formation of rigid structures. Insect cuticle, for example, consists of protein that is extensively cross-linked in this way. (It was only by developing this method of cross-linking that insects were able to develop a light exoskeleton which was nonetheless sufficiently rigid to sustain the stresses of flight. Spiders, which have never achieved flight, did not do so and remain soft-bodied.) Rigid structures may also be obtained by imbedding the protein in a matrix of some material other than protein. The best known example here is bone, in which the matrix is mineral.

Muscle proteins

The principal structural components of muscle cells are bundles of the fibrous proteins actin and myosin. Muscular contraction occurs when the bundles slide between one another, diminishing the size of each muscle cell and thus of the muscle as a whole. This sliding is thought to be the result of a conformational change on the surface of one of the two proteins. The change is almost certainly brought about by the transfer to the protein surface of the chemical energy inherent in a certain bond in the energy-storage compound ATP (see p. 69). Although myosin is in most of its characteristics a typical fibrous structural protein, it is unusual in that it also possesses enzymic activity. The enzymic activity in question is the catalysis of the breakdown of ATP and is clearly connected with the contraction mechanism.

Fibrin and blood clotting

Fibrin is an interesting structural protein. In its native form (fibrinogen) it is a soluble protein of considerable axial ratio (see p. 19). When it is necessary to form a blood clot, three small breaks are made in the peptide chain by an enzyme and the balance of forces stabilizing the fibrous structure is disrupted. A shift occurs to a more globular structure (fibrin) and residues previously buried come to the surface. Many are able to join in hydrophobic and other interactions between molecules and stabilize a great number of inter-molecular links – the analogies with the denaturation process (p. 26) are obvious. Thus the solubility is lowered and a clot forms. The final event is the cross-linking of the molecules by covalent bonds, in particular disulphide bridges.

This phenomenon of activation by removal of a part of the chain is met with elsewhere. It occurs when it would be an embarrassment to have a protein expressing its full function before it was needed. This applies to some hormones and to enzymes which degrade cellular constituents.

Structural proteins of low axial ratio

Globular proteins are sometimes used to solve structural problems, although not so frequently as are the fibrous proteins. An example of the use of globular proteins is the encapsulation of virus nucleic acid into a rigid protective quaternary structure (see p. 55).

Proteins which bind other molecules

We come now to the third and most intensively studied class. These proteins exploit to the full the possibilities of combination of the different amino acid side chains. They do so to produce a site on their surface which has both a specific shape and a specific array of forces. This site will bind a chosen molecule or part of a molecule with great tenacity. This is because it is a perfect fit both in the geometrical sense and in terms of the chosen forces meeting just those parts of the structure to be bound upon which they can best act. Thus the binding is both powerful and reasonably specific, since even a small change in the structure to be bound will be likely to spoil the fit and upset the interplay of forces (see also enzyme specificity, p. 45).

If the sole function of the protein is to bind, as is to some extent the case with the immunoproteins (see below), the matter ends there. In most cases, however, it is necessary not only to bind a molecule but to provide a special environment in which to modify its properties. Haemoglobin (see below) is an example of this; so too are the enzymes (Ch. 5), which enhance the reactivity of the molecules which they bind.

Table 4.1 shows a convenient classification of the better known members of this class.

We propose to discuss just one of these proteins, haemoglobin, in any detail, as an example of all the others. However a few notes on Table 4.1 may be helpful.

The hormones, whether or not they are proteins, are thought to influence metabolic rates by their ability to bind at a specific site.

Table 4.1. *Proteins which bind to other molecules*

Hormones*	Transmit instructions for the control (p. 185) of the levels of metabolic activity
Immunoproteins	Bind to and inactivate foreign materials invading the body
Enzymes	Catalyse biochemical reactions
Carrier proteins	Transport of molecules, ions or electrons from one place to another either within the cell or over greater distances within the organism

* Not all hormones are proteins.

This site may be on a membrane, which has its properties modified as the result of the binding. The rate of passage of metabolites through the membrane might then be significantly altered and this would obviously influence the rate of metabolic activity within the cell or other structure bounded by the membrane. Alternatively, a hormone may bind to a specific enzyme and induce a change in the tertiary structure which alters the catalytic activity. We shall see (Ch. 21) that a change of activity of just one enzyme can frequently influence the rates of a large number of biochemical reactions. Finally it is not impossible that hormones may act on the genetic apparatus and influence the rate of synthesis (p. 175) of an enzyme. To change the amount of an enzyme in a cell is equivalent in many ways to changing its activity and will have similar consequences.

Insulin is a well known protein hormone secreted by certain specialized cells in the pancreas. It is practically the smallest protein known, having only fifty-one amino acids. It has profound effects on, in particular, carbohydrate metabolism. It is thought to act at the membrane.

The immunoglobulins deal with invading substances in the body and are responsible for such phenomena as immunity and transplant rejection. The system works in the following fashion. In some way, the body is able to recognize the presence of certain classes of alien molecules known as antigens. (Antigens are usually macromolecules: protein, nucleic acid, polysaccharide, lipid or a combination of any of these.) Once the presence of the antigen is recognized, the synthesis of the immunoprotein (antibody) begins.

A large number of different antibodies will be produced in response to any one antigen. All will contain a binding site (see above) which will fit and bind to one of the structural features of the antigen that is not found in any of the molecules belonging to the host organism. For instance, suppose that there is an area of the invading antigen with the shape and combination of forces shown in Fig. 4.3.1. Somehow an antibody is selected from the many designs available

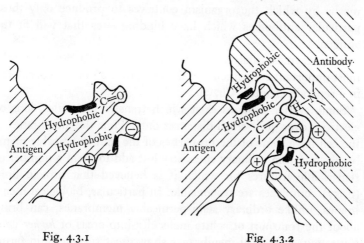

Fig. 4.3.1   Fig. 4.3.2

Fig. 4.3. Antigen–antibody combination.

to the organism with a binding site nearest to that which would provide a perfect fit (Fig. 4.3.2) in both of the senses used on p. 33. Since it appears that an organism has millions of possible structures at its disposal, the best of these is, for most conceivable antigenic features, likely to be a very good fit indeed. If this is so the antibody will bind irreversibly to the antigen.

The antigen thus complexed is likely to lose its capacity to act. A bacterium or tissue transplant covered in antibodies will probably cease to function. Immunization relies on the fact that having once learnt to synthesize an antibody the body is able to produce it very much more quickly should it ever be called upon to do so again. Rapidly multiplying viruses or bacteria will then have less chance of gaining the upper hand before the antibody concentration reaches an adequate level to deal with them.

The protein structure of the antibodies is currently under

investigation. A great deal of information is becoming available and the study of antibody structure is now one of the most promising areas of protein chemistry. It appears that one antibody differs from another only in a limited part of the molecule (which presumably includes the binding site), considerable regions of the structure being more or less invariant. Attention is now turning to the elucidation of the structure of the binding site itself and the means by which the organism contrives to produce only those particular antibodies which have binding sites that will fit the antigen.

The enzymes, catalysts of biochemical reactions, are among the most important constituents of living matter. They are considered in a separate chapter (Ch. 5).

Some carrier proteins exist which transfer molecules, ions or electrons over quite small distances only. Thus the membranes of the cell have some of the properties of the semi-permeable membranes found in ordinary chemistry but add to them a number of remarkable features for which it is believed that protein–lipid (p. 57) complexes are responsible. In particular, biological membranes, unlike ordinary semi-permeable membranes, can bring about the transport of solute molecules into areas of *higher* concentration. Biological membranes show great specificity in terms of which molecules they will transport and in which direction they will transport them. Membranes exist, for example, which will accept D-sugars and reject L-sugars, while nerve membranes will concentrate Na^+ ions on one side and K^+ ions on the other.

The cytochromes (Ch. 8) transport electrons over short distances in the cell, and many enzymes (q.v.) may be said to have short range transport functions as well as catalytic ones. This is because in several metabolic pathways (q.v.) the enzymes concerned are arranged in close and defined proximity to each other in assemblies built into the membrane. The speed of operation of some of the assemblies is such that it appears that there is not enough time for the reactants to diffuse in free solution from one site to another; they must be passed directly from one enzyme surface to another.

Other protein carriers exist to transport a necessary substance from one part of an organism to another quite distant part. These may frequently act as stores, holding the substance carried until it is required. The leading example of such proteins is probably haemoglobin, an oxygen carrier (see below). There are also a number of proteins which carry particular metal ions about the body as well as many which carry other specific small molecules. All rely on the usual covalent and non-covalent forces to provide an area of high affinity for the substance to be carried.

Haemoglobin: biological phenomena explained at the molecular level

In order to show in the greatest possible detail how the principles of Ch. 3 may be applied to explain the activity of proteins we have chosen two examples. One is the enzyme lysozyme, which will be dealt with in the next chapter, the other is haemoglobin. Haemoglobin is chosen to represent the non-enzymic proteins partly because of its medical and biological significance and partly because it presents one of the most rewarding instances of the study of biology in molecular terms.

The site of interaction between oxygen and haemoglobin is a ferrous iron atom. It is prepared for its role in oxygen binding by being held in an environment containing five nitrogen atoms at precisely the correct orientations and distances to allow co-ordination to occur. The positioning of the nitrogen atoms is controlled by having four of them bound together in a rigid ring system of the correct dimensions (the porphyrin ring, see Fig. 4.4). This molecule has several hydrophobic areas, and these and other interactions are used to fix the ring on to a suitably shaped, partly hydrophobic patch on the protein molecule. The nitrogen of a histidine side chain takes up the fifth co-ordinating position. The sixth position, which is called for by the geometry of co-ordinated iron, is found to be particularly suitable for binding oxygen. This binding is strong enough for the carrier function to be fulfilled but not too strong to prevent the oxygen from being given up when required. Carbon monoxide is poisonous because it will bind in the same place but much more strongly and will thus not make room for oxygen when required. This phenomenon is analogous to competitive inhibition in enzymes (p. 46).

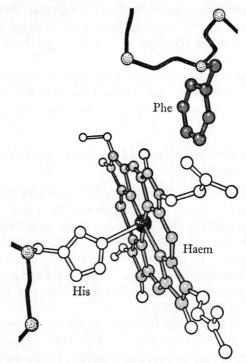

Fig. 4.4. Haem–protein interactions.

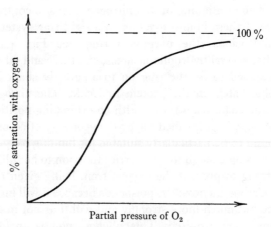

Fig. 4.5. The oxygenation–deoxygenation curve of human haemoglobin

Haemoglobin has been developed by the vertebrates at least into an extremely subtle instrument. It is now possible for us to appreciate how this has been done, since the structure of the molecule has been determined in atomic detail. It is very much in the interests of an animal to have an oxygen carrier which does not have a straight-line relation between the amount of oxygen available (or required) and the percentage oxygenation of the carrier, but a so-called sigmoid relationship (Fig. 4.5). The advantage of the sigmoid curve is that almost the full oxygen capacity of the carrier can be called on with only a small drop in the oxygen concentration. If a straight-line relationship existed then the carrier would still be hanging on to some of its oxygen at very low oxygen levels, and would still be inclined to give up some of its oxygen at very high oxygen levels. This greater efficiency allows the vertebrate cell to exist in an even environment without the great fluctuations in oxygen level which would be necessary to utilize the capacity of a less sophisticated carrier to the full. These advantages must be a contributing factor to the success of the vertebrates as a group in spite of the small number of species (there are roughly 50000 described species of vertebrates out of approximately 1000000 animal species).

The active part of vertebrate haemoglobin is still the porphyrin ring (Fig. 4.4). The improvement in performance is managed by means of a new refinement in the way that the protein side chains influence its environment. What has happened is that four protein molecules have aggregated into a specific quaternary structure (Fig. 4.6). This structure is so arranged that when one ring is oxygenated, the disturbance to the protein structure at that point is transmitted by re-arrangement of the protein side chains to the site of another haem ring. The consequent change in the environment of this second ring is such as to increase its affinity for oxygen. Once one oxygen molecule has gone on to the tetramer another is much more likely to follow. This gives the up-turn in the oxygenation curve which is required to produce the sigmoid shape. The phenomenon of activation (or de-activation) of a site on a quaternary structure by binding of a molecule at a distant site is also exploited in the regulation of the catalytic activities of enzymes (see p. 183).

Haemoglobin has another important biological activity which could easily be overlooked; it is a major source of the buffering power of blood. The dissociation equilibria of its ionizable side chains (many of which project into the solution) control pH very effectively in the same way as the buffering substances used in non-biological chemistry. Even more subtly, the molecule is arranged so that, when the tetrameric structure is disturbed by oxygenation or de-oxygenation, different residues are exposed to compensate for the pH changes which would result from the production of carbonic acid when the oxygen is used up (Fig. 4.6).

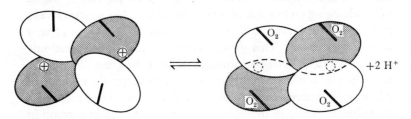

Fig. 4.6. Positively-charged side-chains revealed by the change in shape following deoxygenation of haemoglobin (idealized).

The structure has some remarkable properties and it is clear that even small variations would be likely to have unfavourable results. A number of mutations (p. 151) are known which have given rise in certain individuals to haemoglobins in which one amino acid has been substituted for another. If the change, as is often the case, does result in impairment of function, the individual will suffer from an anaemia. The precise clinical symptoms of this anaemia will depend on the nature and severity of the impairment of function of the molecule. Many anaemias which were first known solely as clinical patterns, sometimes of great complexity, have now been explained completely in molecular terms – a hopeful augury for other diseases. By knowing the location of particular amino-acid substitutions in the haemoglobin of the patients concerned, it is frequently possible to predict the alteration in the properties of the protein from the nature of the substitution. The consequences of this alteration may then be related to the nature of the disease.

In concluding this chapter we must stress that haemoglobin has been selected solely as an *example* of the way in which it is becoming possible to explain quite complex biological phenomena in terms of a few relatively simple ideas. Information of this sort, though possibly a little less complete, exists for a number of other proteins, and much more is likely to become available in the near future.

5 Protein function II – the enzymes

The enzymes are the catalysts of biochemical reactions. No substance with the properties of an enzyme has been found which is not a protein although, of course, many enzymes employ cofactors or prosthetic groups. It is now agreed that should such a substance ever be found, it will not be called an enzyme.

The enzymes are of particular importance as catalysts of most of the reactions of intermediary metabolism. Their catalytic power is prodigious – frequently several orders of magnitude better than the corresponding non-biological catalysts. For example, one molecule of an enzyme which destroys hydrogen peroxide is able to deal with approximately 5 000 000 molecules of H_2O_2 per minute.

Their specificity is equally striking. We have already mentioned (p. 18) a case in which living processes distinguish between different optical isomers, almost the ultimate problem of ordinary preparative chemistry. This is but one example of the ability of enzymes to discriminate between possible substrates, however closely related they may be. Urease is an example of very high specificity. It is very active in catalysing breakdown of urea according to the equation

$$CO(NH_2)_2 + H_2O = CO_2 + 2NH_3.$$

Many compounds exist in which substituents are placed on the NH_2 groups of urea. None are touched by the enzyme.

Another group of enzymes has a somewhat broader specificity, for example, the digestive enzymes catalysing the breakdown of proteins by hydrolysis of the peptide bond. Such enzymes tend to show specificity in that they will only cleave peptide bonds in which certain classes of amino acids form one of the partners. Trypsin will hydrolyse only those peptide bonds formed by the carboxyl groups of the long-chain basic amino acids lysine and arginine (Fig. 5.3.1). Chymotrypsin on the other hand rejects these peptide bonds and attacks those formed by the carboxyl groups of

[42]

a number of other amino acids. Its preference is in fact largely for certain of the hydrophobic amino acids, notably the aromatic ones (Fig. 5.3.2). Finally there may be an almost complete lack of specificity. Enzymes exist which will cleave nearly all peptide bonds while others will hydrolyse nearly all esters, and so on.

It should also be noted that the specificity extends to the products. That is to say in almost all cases a given enzyme will catalyse only one type of reaction with a particular substrate. This does not mean that an enzyme can never catalyse more than one type of reaction when given more than one type of substrate: e.g. proteolytic enzymes usually become esterases when confronted with peptide or amino acid esters.

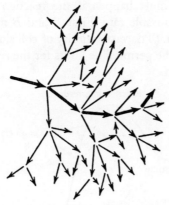

Fig. 5.1. Illustrating the power of a chain of enzymes to select just one of a large number of reaction pathways.

The potential of this great catalytic power coupled to great specificity has extremely important consequences. There is an almost infinite number of reactions which it would be energetically possible for a chemical compound (glucose, say) to undergo. The products of many of these reactions in their turn may each react in a large number of different ways, and so on for their products (Fig. 5.1). As will be seen below, enzymes cannot make energetically unfavourable reactions happen, they simply accelerate reactions that are possible. This acceleration is so great that it amounts to a *selection* process, and in the presence of suitable enzymes the starting material is steered through the maze of reactions as though the ever-multiplying possibilities did not exist. This power of enzymes to *organize* by virtue of the specificity and rapidity of action is even further increased by the way in which many of them are arranged in the cell and by the many subtle mechanisms which provide for the increase or decrease of their activity as required (Ch. 21).

Before we can attempt to explain how enzymes function it is necessary to describe something of how chemical reactions take place. Consider the reaction

$$A + B \rightleftharpoons C + D.$$

Even if the forward reaction is energetically favoured (p. 62) the following things must happen if the reaction is to be seen to proceed at a measurable rate. First A and B must collide in the proper orientation. (The vast majority of collisions in free solution are wasted since the permissible limits for the relative orientations

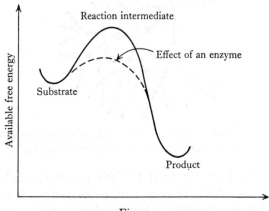

Fig. 5.2

of the particular bonds and groups in the reacting molecules fall within surprisingly narrow limits.) The second requirement is that the 'potential barrier' (Fig. 5.2) to the reaction must be overcome. Because of this barrier, many reactions do not take place even though they are energetically favoured. The barrier may be explained by the fact that while A and B are interacting to give C and D they form a temporary and unstable complex called a reaction intermediate. The free energy state (see p. 61) of this reaction intermediate is often much higher than that of either products or reactants. In such cases the existence of an intermediate, though essential for the progress of the reaction, is at the same time a barrier. In the absence of a catalyst the barrier can be overcome only by a favourable random fluctuation in the energy of the reactants. Heat will increase the magnitude of this fluctuation, which is why heat speeds up chemical reactions.

Enzymes, like any other catalysts, are unable to promote a reaction unless it is energetically favourable. They cannot influence the initial or final energy states of the reactants and so cannot force an otherwise impossible reaction to occur. All they can do is act on both the factors (mentioned above) that control the rate of the reaction and increase the speed with which a reaction, already possible on energetic grounds, will happen.

Those catalysts which act as surfaces, the class to which enzymes belong, can ensure the proper type of collision for reaction by binding A and B in the desirable specific orientation and in close proximity. Furthermore, if the binding is strong, A or B will remain on the enzyme until the arrival of its partner and there will be no need to wait for the relatively rare event of all three mole-cules, reactants and catalysts, colliding. (This is, in effect, increas-ing the local concentration of the reactants.) Thus the first thing an enzyme does is to bind its reactants. The usual forces are employed, and specificity can be ensured by making use of the structural versatility of proteins described in Ch. 3. Fig. 5.3 shows how in two enzymes which degrade proteins the amino acid sequences differ, making sites with different binding proper-ties for the specific types of amino-acid side chain. The analogy has frequently been made between the fitting of the correct substrate of an enzyme into the binding site designed for it and the fit between a key and a lock. This is a static picture and does not allow for the possibility that the enzyme might in some cases wrap itself around the substrate and so form the correct shape of lock only when the key is present. The picture of a lock and key is nonetheless helpful in understanding enzyme specificity.

The mode of action of so-called competitive inhibitors can now be understood. These are compounds which resemble the substrate

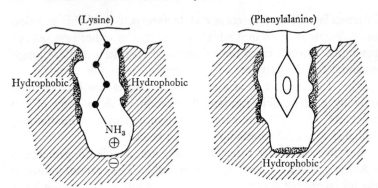

Fig. 5.3.1. The specificity site of trypsin.

Fig. 5.3.2. The specificity site of chymotrypsin.

molecules sufficiently well to form some of the proper interactions with the binding site and yet are not sufficiently similar to take part in the reaction and be released. The wrong key will not turn in the lock, but if it is close enough to the right shape it may jam and cannot then be removed. Thus, consider the enzymic conversion of succinate ion to fumarate ion (Fig. 5.4 and see p. 99).

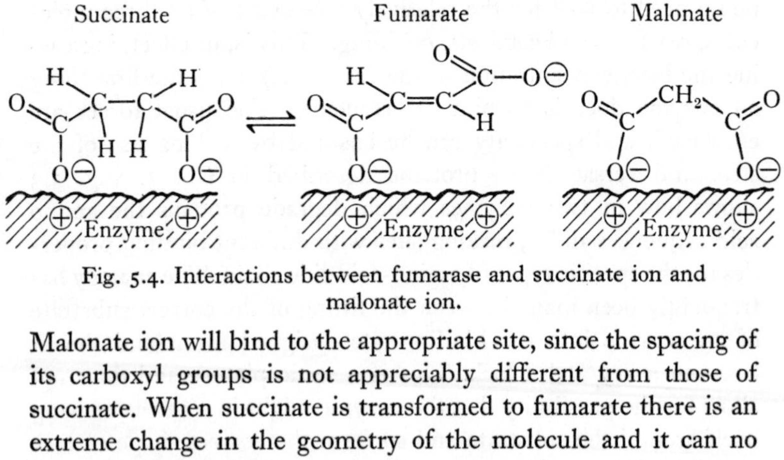

Fig. 5.4. Interactions between fumarase and succinate ion and malonate ion.

Malonate ion will bind to the appropriate site, since the spacing of its carboxyl groups is not appreciably different from those of succinate. When succinate is transformed to fumarate there is an extreme change in the geometry of the molecule and it can no longer be bound. Malonate clearly cannot undergo such a change and so it continues to occupy the binding site. While it remains the enzyme is unable to accept any further molecules. Malonate is therefore a powerful inhibitor of the enzyme.

As an example of the utility of protein binding sites in raising the effective substrate concentration, we may consider the class of reactions which involve hydrolysis and in particular those in which protons are the active agent. The rate of such reactions will be dramatically enhanced if proton binding takes place. The concentration of protons at physiological pH is, of course, quite small: 10^{-7} moles per litre at pH 7 by definition. (The volume of the cell is so small that this will mean only a hundred or so protons per cell.) The storage of a proton in a position in which it may be used in a hydrolytic reaction is equivalent to a drastic lowering of the pH. It does not, however, have all the undesirable consequences to the cell of a general increase in hydrogen ion concentration.

Students of thermodynamics will recognize that increasing the number of useful collisions represents an entropic contribution to the lowering of the free energy of the potential barrier. In addition enzymes may lower the potential barrier more directly. If some of the side chains on the protein surface are able to interact with the reaction intermediate they may stabilize it. The stabilization of a structure is synonymous with the lowering of its free energy state. It can be shown that the probability of random thermal fluctuation sending the reactants over the potential barrier to the product side increases exponentially with the lowering of the barrier; the reaction is accordingly speeded up in a dramatic way (Fig. 5.2). Fig. 5.5.2 shows something of the proposed stabilization of a reaction intermediate in the case of the enzyme lysozyme, which hydrolyses polysaccharide chains in bacterial cell walls. We see in the figure the chain having already been cleaved by a proton donated from the carboxylic residue marked 35. The resulting positively charged intermediate would be grossly unstable in free solution but it is given some stability here. This is due to the proximity of the negative charge of the carboxylic acid group No. 52 and secondly to interactions with the suitably shaped surface of the enzyme molecule which hold the ring in the deformed configuration necessitated by the geometry of the intermediate.

This figure also helps to illustrate one other feature of the relation between enzyme structure and function to which we have frequently referred. This is the placing of reactive groups in unusual environments which confer on them unusual properties.

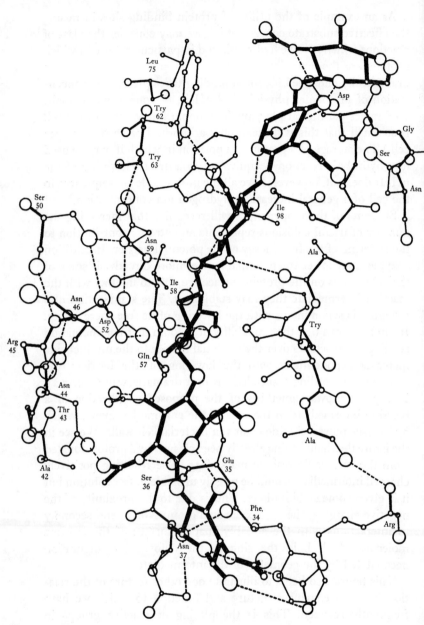

Fig. 5.5.1 (*See facing page*).

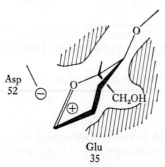

Fig. 5.5.2. Idealized picture of the stabilization of the transitional inter-mediate formed during the hydrolysis of the bond between the second and third sugars from the bottom in Fig. 5.5.1. Contacts with amino acid side chains stabilize the otherwise improbable conformation of the sugar ring. The negative charge of Asp 52 stabilizes the positively charged inter-mediate by means of an ionic bond.

Carboxylic acid residue 35 is in an exceptionally hydrophobic environment. In such an environment it is very much more difficult for the group to ionize, in other words if it finds a proton it will keep it until it donates it in the reaction. This then is a very strong proton binding site. Carboxylic acid residue 52 on the other hand is in a very hydrophilic environment and under these circumstances ionization is favoured. Its environment therefore helps it to remain charged, so that it is able to assist in stabilizing the intermediate structure in the way referred to above.

Each organism probably contains many thousands of enzymes. It seems likely that with the addition of co-factors and prosthetic groups (which will be referred to frequently later on), all rely on something like the above principles for their remarkable properties. The enzymes give the chemical activities of the living state some of the most striking of their unusual features.

Fig. 5.5.1. A computer-drawn diagram showing the interaction between a hexasaccharide substrate (bold lines) and the amino acid side chains of the binding site of lysozyme. Oxygen atoms are shown as large circles and nitrogen atoms by slightly smaller ones. Hydrogen atoms are omitted. Hydrogen bonds are represented by dotted lines. Note the complexity and specificity of the interaction. Some of the individual bonds are easily traced, e.g. the hydrogen bond between the $CH_3.CO.NH$–substituent on the second sugar ring from the bottom and the side chain of asparagine 44. Others are less easy to follow without a stereoscopic drawing.

6 Nucleic acids, polysaccharides and lipids

Nucleic acids

The impression will have been formed in the last chapter that the proteins have an important part to play in every distinctive activity of living processes save one. The apparent omission was reproduction. Even here proteins do have their part to play but the molecules most directly involved are the nucleic acids. This indeed is their only major function and this fact explains why they possess a much narrower range of residue than the proteins. It also explains the relatively small number of types of nucleic acid.

The most common bases found in nucleic acids are shown in Table 6.1. They are linked by a glycosidic bond (see p. 7) to a sugar – D-ribose or D-deoxyribose depending on whether the nucleic acid is ribonucleic acid (RNA) or deoxyribonucleic acid (DNA). The resulting base–sugar compound is called a nucleoside. When an —OH group in the sugar of a nucleoside is phosphorylated, the compound becomes a nucleotide. The nucleotides are linked in a specific order to form the nucleic acid by means of the phosphodiester linkage, as described previously (p. 5).

Once again the full range of forces comes into play in the stabilization of the three-dimensional structure. The hydrogen bond receives most attention because of the remarkable way in which the bases, when in the same plane, fit into one another in two sets of complementary pairs stabilized by hydrogen bonds (Fig. 6.1).

The well known double helical structure (Fig. 6.2) involves such coplanar pairs of bases. It will be seen that in each strand the bases are stacked upon each other like the steps in a spiral staircase. This structure is only stable when these pairs are complementary in the sense used above. Thus the base sequence of one strand can predetermine the base sequence of the other (see Fig. on p. 160). It is this device (which relies on hydrogen bonding) upon which the self-replication of the molecule and the control of the sequences of the other molecules depends (see Ch. 18).

Table 6.1 *Bases and sugars in nucleic acids. (Note the various systems for numbering the atoms in the rings)*

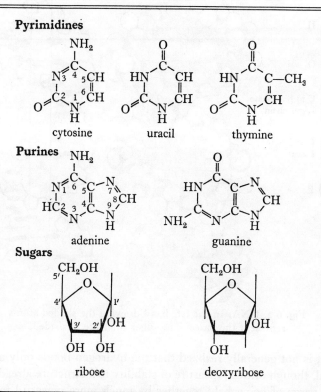

Pyrimidines

cytosine uracil thymine

Purines

adenine guanine

Sugars

ribose deoxyribose

Fig. 6.1. Base pairing.

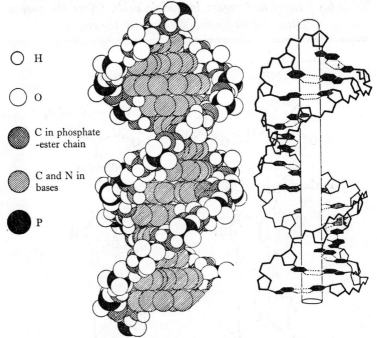

H

O

**C in phosphate
-ester chain**

**C and N in
bases**

P

Fig. 6.2. DNA. In the left-hand drawing the shaded atoms
represent the bases. The sugar rings are unshaded.

It is not generally realized that the hydrogen bonds only add a
final, though critical, measure of stability to a structure already on
the verge of being held together by much more powerful forces.
Of the non-covalent forces previously mentioned hydrophobic and
dipole interactions both stabilize the stacking of the bases in each
strand and push the two strands together. It is in fact this squeezing
action that forces water molecules out of the centre of the assembly
and leaves a clear field for the hydrogen bonds to form (Fig. 6.3).

The double helical structure is characteristic of DNA. A closely
analogous double helix exists for RNA wherever complementary
base pairs are found. This structure is somewhat different because
of the intrusion of the bulk of the extra oxygen atom at position 2'
of the ribose into a crowded part of the structure (Fig. 2.2).
Fig. 6.4 shows that completely double-stranded RNA molecules
do exist but the structure is more frequently found only in stretches

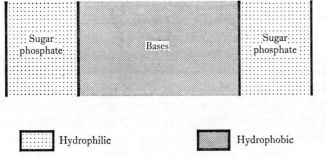

| :::Sugar::: :phosphate: | Bases | :Sugar: :phosphate: |

☐ Hydrophilic   ■ Hydrophobic

Fig. 6.3. Hydrophobic and hydrophilic regions in DNA.

Fig. 6.4. Double-stranded RNA. In contrast to DNA, the bases are neither parallel to one another nor perpendicular to the axis of the helix.

of the molecule where complementary sequences happened to occur in positions that enable them to come together. These double-stranded regions are separated by single-stranded regions where no complementary sequence is available (Fig. 20.1). Such a structure will be in violent agitation through random molecular bombardment and is by no means static. Divalent cations such as magnesium will help to make the various parts of the structure cohere by acting as a positively charged adhesive between two negatively charged phosphate groups on different parts of the structure. No other types of cross-linking are known, in contrast with the proteins and polysaccharides.

DNA is found outside cells as single- or double-stranded virus nucleic acid (see p. 55) 10^3 to 10^5 bases in length. Inside the cell the best known form is that found in the nucleus in the chromosome, a complex between double-helical DNA and predominantly basic proteins. It acts as the library of genetic information for the cell (see Ch. 18). It has a very large chain length, hundreds of thousands of bases. The DNA of a bacterium is 1.4 millimetres long and were it not for its thinness this would be visible to the naked eye.

RNA is found outside cells as virus nucleic acid. It is most usually single stranded and is 10^3 to 10^4 bases in length. Inside the cell there are three types. The first is *messenger RNA* (mRNA) which carries the genetic message (p. 166) from the nucleus to the site of translation. This is a single-stranded molecule of a length of the order of 10^3 to 10^4 bases. *Ribosomal RNA* (rRNA) occurs in ribosomes, which are the subcellular components involved in the translation of the genetic message (p. 165). There are three nucleic acid components in the ribosome. One is of about 3×10^3 bases, one is half that length and one is just over 10^2 bases. Any double-stranded structure in them is of the type discussed on p. 53.

Transfer RNAs (tRNA) carry the amino acid residues to the ribosome to be incorporated into protein in an order dictated by the hydrogen-bonding properties of the nucleotide sequence of the messenger (p. 171). Each amino acid has at least one species of tRNA molecule of its own with about 70–80 bases in a characteristic sequence. A more detailed description of the structure of tRNAs occurs on p. 168, where it is possible to discuss the structure more explicitly in terms of the details of the way in which the molecule functions. For the moment the reader should glance briefly at Fig. 20.1, p. 168, and note the alternation of the double and single-stranded regions, which were referred to above.

Nucleic acids are frequently found in combination with basic proteins (histones, nucleoproteins, ribosomal proteins, etc.). The precise mode of action of these proteins is obscure, but the basic proteins associated with DNA may have some role in the control of the expression of genetic information (p. 181), while the ribosomal proteins contribute an essential element to the proper structure of the ribosome. Ionic forces probably mediate in the interactions between nucleic acid and basic proteins. Many of the basic proteins found in conjunction with nucleic acid have as much as 80% of their side chains drawn from the basic group in category 2 in Table 3.1, p. 17. Many ionic bonds may then form between the positively charged side chains and the negatively charged phosphate groups of the sugar phosphate backbone.

Another class of proteins found in association with nucleic acid do not rely so much on ionic bonds. These are the virus proteins which are so constructed as to have a strong quaternary structure. This encloses and protects the nucleic acid (Fig. 6.5).

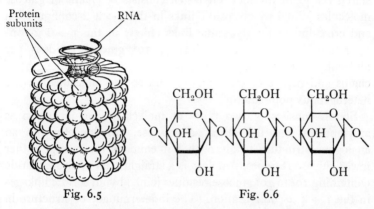

Fig. 6.5   Fig. 6.6

Fig. 6.5. Quaternary structure in a cylindrical virus.
Fig. 6.6. Cellulose. Compare with the $1 \rightarrow 4\alpha$ configuration shown in Fig. 2.3.

Polysaccharides

Certain D-hexoses (especially glucose and its derivatives) and D-pentoses can, as already mentioned, undergo condensation via the glycosidic bond. The molecules formed, which may be of considerable size, normally perform structural duties or act as sugar reserves. Cellulose is the best known example of a poly-saccharide with considerable mechanical strength. It is formed of unbranched chains of glucose residues joined by the glycosidic linkage in the $1 \rightarrow 4$ β-configuration (see Fig. 6.6). Such chains may pack closely together and form a strong bundle with many possibilities for hydrogen bonding between them. For greater strength, in wood for example, extensive cross-linking by aromatic free-radical reactions may occur, in a way somewhat reminiscent of the proteins (p. 31). Many other sugar polymers are known with similar functions to those of cellulose. When other sugars are used, particularly those in which some of the hydroxyl groups have been chemically modified, molecules can be formed that do not have rigidity. They may then, for example, form solutions with great value as lubricants in the joints of animals, etc.

Both animal and plant tissues make use of polysaccharides as food storage materials. Glycogen typifies the animal product, starch the plant product. Glycogen consists of chains of glucose molecules joined by glycosidic links in the $1 \to 4$ α-configuration and cross-linked by glycosidic links chiefly in the $1 \to 6$ α-configuration. There are approximately 10^4 glucose residues per molecule. There is only approximate control over the lengths of chains and degree of cross-linking and so the molecules form a heterogenous population.

Starch consists of two types of molecule. One is known as amylopectin and the other as amylose. Amylopectin has an analogous structure to glycogen but contains on average rather fewer glucose residues. Amylose is a straight chain polysaccharide containing 10^2 to 10^3 glucose residues joined by glycosidic linkages in the $1 \to 4$ α-configuration. (The indeterminacy of structure in food storage polysaccharides is not paralleled by any indeterminacy in food storage proteins which appear to have sequences as rigidly defined as any other protein. This difference is probably because certain quite small divergences from the correct structure could render such proteins useless (for example by their being made insoluble) in a way which is not easy to imagine for polysaccharides.)

Storage and structural polysaccharides are widely distributed among micro-organisms, plants and animals. They are frequently found in combination with proteins (where they are called glycoproteins or mucopolysaccharides) and with lipids. Such combined materials have many important structural and mechanical functions.

Lipids

As mentioned in Ch. 2 there exists another class of biological molecules which form structures of great size. These are the lipids, which are derivatives of fatty acids. Very many lipids, particularly those of relevance to a consideration of macromolecules, are derived from the esterification by fatty acids of a few polyhydric alcohols. Of these glycerol is the most common (see Fig. 6.7). In the figure R_1 and R_2 are usually long-chain fatty acids and X can be one of a variety of groups, either neutral or charged, polar or non-polar. Here there is no combination of small molecules into long chains by condensation (the exclusion of water) and thus no questions of sequence or the control of sequence. Even with the longest of the fatty-acid substituents lipids hardly approach in molecular weight the smallest of the true macromolecules. They

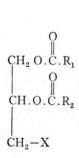

Fig. 6.7

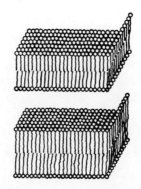

Fig. 6.8. A typical lipid aggregate – in this case a paired-planar structure.

are mentioned here because lipids can use the range of non-covalent forces (q.v.) to associate into very large structures indeed (note the similarities of the individual elements to those in Fig. 2.5). These structures are intimately associated with the other types of macromolecule (see below).

Membranes

It is probable that few of the processes described in this book take place in anything approaching free solution. Very many of the proteins most active in metabolism are bound to the various membrane structures formed by lipid aggregation. These include the limiting membrane of the cell itself, the endoplasmic reticulum and the cristae of the mitochondrion (see Frontispiece). In many cases it is thought that this binding exists for other purposes besides simply fixing the protein to the nearest convenient surface. Many systems of biologically active proteins require the close proximity of a number of different protein molecules in a defined spatial relationship to each other (see p. 133). The lipid membrane may have a specific role to play in the control of the formation of these assemblies.

Section II

METABOLISM

7 Energy and biochemical reactions

It is every day experience that all processes can be divided into two categories, those that tend to occur by themselves and those that must be driven by the expenditure of energy. Water running downhill is an example of the first class, a vehicle moving uphill is an example of the second. Furthermore we know that it is processes of the first class that yield the energy to drive those of the second (a water wheel may be used to pull a trolley uphill). This division applies just as much at the molecular level and is of great importance in understanding the design of living systems. In particular, by analogy with the mechanical examples cited *it is possible to couple the chemical energy yielded by a process of the first type to drive a chemical or physical process of the second type.*

Living matter depends on this possibility. Organisms exploit it to bring about syntheses of compounds and structures which would not otherwise be feasible and also in order to couple chemical energy with mechanical work. (Examples of coupling between chemical and mechanical processes are muscular contraction and the forced transport of solutes into areas of higher concentration.) Just as the wholesale exploitation of energy coupling is a characteristic of life at the molecular level, so the consequences of that coupling – the ability to grow, to move and to organize – are the characteristics by which we recognize life at the macroscopic level.

To assess the importance of any particular process in the economy of the living state we need to know in which category it lies and to have some idea of the amount of energy that could be obtained from it or that would be needed for it. There exists a useful thermodynamic quantity to help us. Consider a process $A \rightleftharpoons B$ taking place at constant temperature and pressure – the conditions under which living matter normally operates. The maximum useful energy produced or consumed is known as ΔG, the change in the so-called Gibbs free energy of the reacting

[61]

system. By convention, if ΔG has a negative value when going from A to B, energy may be obtained from the process and the process $A \to B$ is of the first, energy-yielding, category. If the value is positive then $A \to B$ is of the second, energy-requiring, category.

Clearly if ΔG is negative for $A \to B$ then, as the process is reversible, it will be positive for $B \to A$. This convention is not so illogical as it may seem: if we obtain energy (negative ΔG) from the process the system itself must lose it.

Thus the sign of ΔG tells us in which category the process lies. Our second requirement, to know how much energy is involved, is met by its magnitude. The greater the negative magnitude of ΔG, the more energy is to be had. The greater the positive magnitude, the more is needed.

It must not be thought that if ΔG is negative the process will always and instantaneously occur. To return to the water analogy, consider a reservoir a certain height above the surrounding countryside. There is clearly a strong tendency ('high negative ΔG') for the water to run down. Whether it actually does so or not, and the rate if it does, depends on factors unrelated to this value. If the dam is effective the water will never run down; if there is an infinitesimal hole it may do so but at an imperceptible rate; if the dam is destroyed it will do so with great violence.

Thus returning to chemical reactions we can speak of a reaction being energetically favoured (negative ΔG) but kinetically hindered (see p. 44). The violent combustion of our bodies in air has a large negative ΔG. We survive because at room temperature there is kinetic hindrance of the process – the so-called 'potential barrier' (see Fig. 5.2).

The value of ΔG must depend on the quantities of the components present. A cupful of water descending from the reservoir could not generate the same amount of energy as a million gallons.

This dependence on concentration leads to a very valuable relationship between ΔG and other parameters of chemical reactions. Consider the reversible reaction $A \rightleftharpoons B$ in which $A \to B$ is energetically favoured over $B \to A$. If we start with equal concentrations of A and B the overall reaction will proceed from left to right and the concentration of B will build up while that of A will decrease. The effective ΔG of $B \to A$ will increase in negative value as the concentration of B increases, and the effective ΔG of $A \rightarrow B$ will fall in negative value as the concentration of A falls. Eventually the concentrations will reach a point at which the ΔG's for the forward and backward reactions are equal. At this point no further change in the overall concentrations of A and B will occur.

What we have described, of course, is the approach to equilibrium of the reversible reaction $A \rightleftharpoons B$. It should, therefore, not be surprising to find a relationship between ΔG and the equilibrium constant K, since the latter quantity simply expresses the ratio between the products of the forward and of the backward reactions.

The relationship is

$$\Delta G^0 = -RT \log_e K, \tag{1}$$

where ΔG^0 is the so-called standard change in the Gibbs free energy of the forward reaction at $25\ ^\circ$C. This means that the effective concentration of all reactants is taken as molar (gases at one atmosphere). R is the Boltzmann constant, T is the absolute temperature.

For numerical calculations this expression reduces to

$$\Delta G^0 = -1\,364 \log_{10} K. \tag{2}$$

This equation considerably extends the uses to which a knowledge of the value of ΔG^0 can be put, since it allows us to calculate the extent to which a reaction will proceed as well as the overall direction and the possible yield of chemical energy. Alternatively, if the equilibrium constant is already determined, we may at once calculate ΔG^0.

Values of ΔG^0 are not known for all biochemical reactions. Even where values are known many have been determined in free solution and not under conditions realistically reproducing those in the cell. Nonetheless the values that are available probably do not misrepresent the position too seriously. They have proved so useful that we may expect further and more accurate values of ΔG^0 to become available as time goes on.

Energy in metabolic reactions

To illustrate the use of the ideas contained in the previous section, let us consider some biochemical reactions drawn from later chapters. First a reaction in which $\Delta G^{0'}$† is fairly near zero.

Glucose-6-phosphate $\rightleftharpoons$ fructose-6-phosphate

$$\Delta G^{0'} = +500 \text{ calories/mole} \tag{3}$$

is a reaction of some importance in both the synthesis and degradation of carbohydrates (pp. 128 and 90). From equation (2)

$$\log_{10} K = \tfrac{500}{1364} = -0.367 = \bar{1}.633,$$

$$K = \text{antilog } \bar{1}.633 = 0.43.$$

Thus at equilibrium at pH 7 there will be 0.43 times as much fructose-6-phosphate as glucose-6-phosphate. If there is 1 micromole of glucose-6-phosphate present there will be 0.43 micromoles of fructose-6-phosphate at equilibrium.

There are important consequences of the fact that the equilibrium constant is near to one. Biochemical reactions usually exist to bring about the net formation of a compound which may be required either for itself or as the starting material of a further process (Ch. 9). Reactions with equilibrium constants near unity can, in distinction to some that we shall discuss below, bring about a useful net formation of product in either direction. If one were to add 14.3 micromoles of fructose-6-phosphate to the system, a net formation of 10 micromoles of glucose-6-phosphate would result. If one were to add 14.3 micromoles of glucose-6-phosphate there would be a net synthesis of 4.3 micromoles of fructose-6-phosphate, a reasonable amount, even though the reaction in this direction is slightly disfavoured energetically. Thus the reaction may be used in degradation of carbohydrate, which is found (p. 90) to require the reaction from left to right, and in synthesis, which is found (p. 128) to require the reaction from right to left.

Let us turn to another reaction in carbohydrate metabolism, the

† $\Delta G^{0'}$ is ΔG^0 corrected to pH 7 rather than pH 0 (molarity of H^+ = 1) as would be demanded for those reactions involving H^+ by the definition of ΔG^0. pH 7 is, of course, much more realistic.

equilibrium between dihydroxyacetone phosphate and 3-phospho-glyceraldehyde.

$$
\begin{array}{ccc}
CH_2OH & CHO & \\
| & | & \\
CHOH & \rightleftharpoons\ CHOH & \qquad \Delta G^{0\prime} = +\ 1830\ \text{calories/mole.} \qquad (4)\\
| & | & \\
CH_2O\textcircled{P} & CH_2O\textcircled{P} &
\end{array}
$$

The equilibrium constant is

$$\text{antilog} - \tfrac{1830}{1364} = \text{antilog}\ \bar{2}.66,\ \text{i.e. about}\ \tfrac{1}{22}.$$

At equilibrium there will be 22 micromoles of dihydroxyacetone phosphate to one of 3-phosphoglyceraldehyde. Now we shall see (p. 91) that carbohydrate breakdown requires the reaction from left to right. Although more extreme examples are to come, the 'wrong' product seems even here to be very much favoured by the equilibrium constant. How can the process be used for a significant net formation of 3-phosphoglyceraldehyde and thus a significant breakdown of carbohydrate? The answer is that an enzyme system exists which is ready to take such 3-phosphoglyceraldehyde as there is and convert it very rapidly to the next product in the chain of reactions leading to carbohydrate breakdown. In an attempt to restore equilibrium more 3-phosphoglyceraldehyde will be formed to replace it. So long as the enzyme system continues to tap off the 3-phosphoglyceraldehyde as it is formed, a useful net formation of product will occur. Thus, virtually any quantity of dihydroxyacetone phosphate can be transformed to 3-phosphoglyceraldehyde.

It is necessary, of course, that the overall equilibrium constant of the process which removes 3-phosphoglyceraldehyde should favour removal rather than synthesis. Putting this in free energy terms, the $\Delta G^{0\prime}$ of the removal process must be sufficiently negative overall to overcome the positive $\Delta G^{0\prime}$ of reaction (4).

Another most important example of the use of this 'tapping off' process to control the direction of net formation of product is given on p. 104.

Now for a reaction with a very high positive value of $\Delta G^{0\prime}$:

$$\text{Ribulose-5-phosphate} + \text{phosphate} \rightleftharpoons \text{ribulose-1; 5-diphosphate} \qquad (5)$$

$$\Delta G^{0\prime} = +\ 3200\ \text{calories/mole,}$$

$$K = \text{antilog}\ -\tfrac{3200}{1364} = 4.6 \times 10^{-3}.$$

Ribulose-1;5-diphosphate is to be used for the synthesis of compounds such as carbohydrates. Because of this one cannot, by analogy with the removal of 3-phosphoglyceraldehyde in equation (4), achieve a net production of diphosphate by tapping it off to a lower energy state than the start of the process. This is because the available free energy of any useful products formed from the diphosphate remains too high.

Furthermore, the equilibrium constant makes it impractical to achieve a useful net formation of the diphosphate by adding reactants on the left of the equation as can be done in the case of equation (4). How then is this reaction and the many others like it, equally vital to living matter and equally unfavourable on energetic grounds, to be brought about? The answer is as we saw on p. 61 that processes may be coupled. Thus the problem is solved by *coupling* the processes of high positive ΔG to those of high negative ΔG.

Before we can consider the systems used for coupling we must round off this survey of the energetics of biochemical reactions by quoting an example of a reaction with a high negative ΔG, the equilibrium between phospho-*enol*pyruvate and pyruvate (see p. 92).

$$
\begin{array}{ccc}
CH_2 & & CH_3 \\
\| & & | \\
CO\textcircled{P} + H_2O \rightleftharpoons & C{=}O + P_i \\
| & & | \\
COOH & & COOH
\end{array}
\tag{6}
$$

$$\Delta G^{0\prime} = -13\,000 \text{ calories/mole.}$$

This is a step in the degradation of carbohydrate and it is clear that there is at least the possibility of a useful energy yield in this process which might be coupled to those of high positive ΔG. This possibility, as we shall see later, is made a reality. For the moment we note that the equilibrium constant is antilog $\frac{13\,000}{1364} = 3.7 \times 10^9$ in favour of the forward (left to right) reaction.

This example helps to explain the misuse so often made in biochemistry of the terms 'reversible' and 'irreversible' as applied to metabolic reactions. These terms are not to be taken to refer to the reversibility or otherwise of the *mechanism* of the reaction. They refer only to the ease or otherwise of achieving net formation of product in both directions. The reaction fructose-6-phosphate $\rightleftharpoons$ glucose-6-phosphate was an example of a 'reversible' reaction in this sense. The reaction phospho-*enol*pyruvate $\rightleftharpoons$ pyruvate (note the double arrows) is an 'irreversible' reaction in the sense that one would not seek (without some external energy source) to use the process for the net synthesis of phospho-*enol*pyruvate. We propose to indicate reactions of this sort by the symbol $\rightleftharpoons$. By contrast, reactions in which net formation of product may occur in either direction are indicated by the normal symbol, $\rightleftharpoons$. There is obviously no sharp boundary between the two types of reaction. We have tended to be sparing in the use of $\rightleftharpoons$, reserving it for particularly clear cut cases. In addition, if the energetics of reaction are in doubt (see p. 63), we have used $\rightleftharpoons$.

Energy sources and energy coupling

A chemical reaction is in a sense solely concerned with electrons. If the electrons that maintain the structure of chemical compounds shift permanently from one configuration to another, we say that a chemical reaction has taken place. If the potential energy of the first configuration is higher than that of the second (more negative G) then the reaction could be made to yield energy. If the reverse is true it will require energy. If we imagine ourselves designing a living organism we must search for a source of energy to couple to its essential energy-requiring processes. We are thus really looking for electrons in a state of high potential energy (high negative ΔG) and a sink of lower potential energy (less negative ΔG) into which we can put them.

Save for a few trivial exceptions there is only one *primary source* of electrons at a high potential, able to fall to a lower, which is suitable for coupling to the energy need of living organisms.

This is the photosynthetic process (Ch. 14), carried out in a 'solid state' device found in plant cells which is analogous to but much more sophisticated than a transistor. The device, which is known as the chloroplast, consists of an assembly of proteins (Ch. 3), prosthetic groups (p. 28) and lipids. A quantum of light is absorbed and its energy ($\Delta G^{0'}$ of about 20000 calories per g. equivalent of red quanta) is given up in promoting an electron to a higher potential energy state. The electron then falls back to its original state. In a way which is not fully understood, the energy of the fall is coupled to the synthesis of certain so-called '*high energy*' *compounds*. These compounds, which are discussed below, can undergo energy-yielding reactions, the energy of which can be coupled to energy-requiring reactions. They can be regarded as a *secondary source* of free energy. By the use of this small group of 'high energy' compounds a large number of molecules can be made which would otherwise be energetically disfavoured to a greater or lesser extent; for example carbohydrate and fat (Chs. 15 and 16). These products then form a *tertiary source* of electrons at a useful potential energy level. The plants themselves use photosynthetically produced compounds as food reserves (p. 56). Animals obtain these compounds either by consuming vegetable matter directly or, if they are carnivorous, via the food chain.

Given that we have a source of electrons at a high configurational energy, we must now look for a sink at a lower energy state into which they may fall, so that the energy of the fall may be coupled to endergonic processes. Simple processes like

$$Fe^{2+} \rightleftharpoons Fe^{3+} + e^- \tag{7}$$

remind us that to donate an electron is to be oxidized. Although other molecules exist which are capable of acting as acceptor, molecular oxygen owes its almost unrivalled ability to act as an electron acceptor to certain peculiarities of the way in which its electron shells are filled.

Oxidative metabolism is, therefore, a very important source of energy. Oxidation reactions of the type

$$AH_2 + \tfrac{1}{2}O_2 \rightleftharpoons A + H_2O \tag{8}$$

can easily yield a $\Delta G^{0'}$ of 30 000 to 50 000 calories/mole.† This would be more than enough to drive the endergonic processes of the cell, reaction (5) for example. Cells that utilize oxidative reactions as an energy source also have a 'solid state' device, this

† Readers may be accustomed to treating oxidation reactions differently from all others and expressing the energy levels in terms of standard electrode potentials. These potentials are directly convertible to ΔG^0 (a change in standard potential of $+1$ volt for 1 g. equivalent of electrons at pH 7 corresponds to a $\Delta G^{0'}$ of 23 000 calories/mole). In order to stress the unity of all energy-yielding reactions the one scale, ΔG^0, will be used throughout this book.

time called the mitochondrion (Ch. 8, Frontispiece), to trap this energy. The components of the mitochondrion are, as far as the different functions permit, similar to those of the chloroplast. The 'high energy' compounds produced are identical.

It is now time to describe the high energy compounds and the way in which they are used in energy coupling.

High energy compounds

Both the photosynthetic and oxidative processes produce the molecule adenosine triphosphate (ATP). The structure is given in

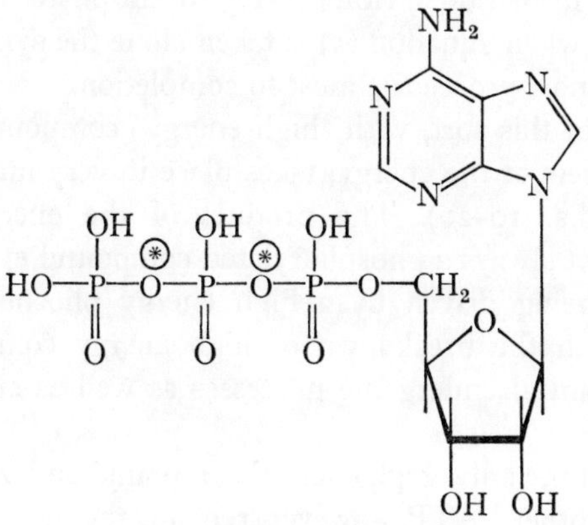

Fig. 7.1. ATP.

Fig. 7.1. It is a property of the phosphodiester bond (marked with a star in the figure) that it has a considerable free energy of hydrolysis.

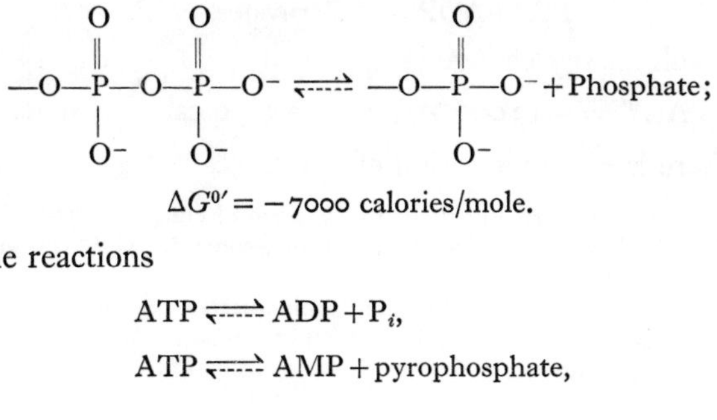

$$\Delta G^{0\prime} = -7000 \text{ calories/mole.}$$

Thus the reactions

$$\text{ATP} \rightleftharpoons \text{ADP} + \text{P}_i, \tag{9}$$

$$\text{ATP} \rightleftharpoons \text{AMP} + \text{pyrophosphate,} \tag{10}$$

have a $\Delta G^{0\prime}$ of -7400 and -7600 calories/mole respectively.†
During the photosynthetic or oxidative synthesis of ATP the
energy of the falling electron is somehow used to drive reactions
such as (9) from right to left. The resulting ATP can now move
to a site in the cell at which, for example, reaction (5) is to proceed.
Combining reactions (5) and (9), we have

ribulose-5-phosphate + ATP $\rightleftharpoons$ ribulose 1; 5-diphosphate
$$+ \text{ADP} \quad (11)$$
$$\Delta G^{0\prime} = +3200 \,(\text{equation } 5) - 7400 \,(\text{equation } 9)$$
$$= -4200.$$

The equilibrium constant is now antilog $\frac{4200}{1364} = 1.2 \times 10^3$ in
favour of the formation of ribulose-1;5-diphosphate. In contrast to
the situation when equation (5) is taken alone the synthesis of the
diphosphate now proceeds almost to completion.

Coupling of this sort, with 'high energy' compounds as intermediate carriers of the energy, takes place in very many chemical
reactions (Chs. 10–20). The product of the energy-requiring
reaction is not always a phosphorylated compound even when the
reaction is being driven by a high energy phosphate (p. 155).
Coupling with the breakdown of high energy compounds also
drives mechanical endergonic processes as well as chemical ones
(p. 61).

ATP is not the only 'high energy' compound and Table 7.1 lists
a number of others. ATP, however, remains the principal medium
of energy exchange.

Phospho-*enol*pyruvate (6) can now be seen as an example of a
high energy‡ compound. It is not used as such directly, but indirectly, through the manufacture of ATP.

$$\text{PEP} + \text{ADP} \rightleftharpoons \text{Pyruvate} + \text{ATP}. \quad (12)$$

The overall energy change is

$$\Delta G^{0\prime} = -13\,000 + 7400 = -5600 \text{ calories/mole}.$$

Thus there is a net formation of ATP.

† ATP and ADP normally occur in the form of complexes with Mg^{2+}. The
values of $\Delta G^{0\prime}$ given for reactions (9) and (10) assume the Mg^{2+} complexes to be
the reacting species.
‡ 'High energy' is a despised term in some quarters and does indeed have its
drawbacks. However, it is a useful shorthand term and as long as it is used with
care it is worthy to be retained.

Table 7.1. *High energy compounds*

Type	Common examples	See page
Pyrophosphates $\begin{matrix} \quad OH \quad\quad OH \\ \quad \mid \quad\quad\quad \mid \\ -P-O-P- \\ \quad \parallel \quad\quad\quad \parallel \\ \quad O \quad\quad\quad O \end{matrix}$	ATP, ADP; other nucleoside di- and triphosphates	69, 81, 116, etc
Acyl phosphates $\begin{matrix} R-C-O-\textcircled{P} \\ \parallel \\ O \end{matrix}$	1; 3-diphosphoglyceric acid	91
Enol phosphates $\begin{matrix} R-C-O-\textcircled{P} \\ \parallel \\ CH_2 \end{matrix}$	Phospho-*enol*pyruvic acid	92
Thioesters $\begin{matrix} R-C-S-R' \\ \parallel \\ O \end{matrix}$	Acetyl CoA	97
Guanidine phosphates $\begin{matrix} R-C-NH-\textcircled{P} \\ \parallel \\ NH_2 \\ \oplus \end{matrix}$	Creatine phosphate R = CH_3 used as energy store $\quad\quad\quad\mid\quad$ in muscle $\quad\quad\quad N-$ $\quad\quad\quad\mid$ $\quad\quad\quad CH_2$ $\quad\quad\quad\mid$ $\quad\quad\quad COOH$	—

This example also serves to point out that there are other means of producing ATP than photosynthetic or oxidative phosphorylation. This third type of process, in which the phosphorylation arises as part of the mechanism of a metabolic reaction, is called substrate-level phosphorylation. Although substrate-level phosphorylations are not of such general significance as the other two types, important examples do exist (pp. 92 and 99).

It may now be realized that we have had for some time an example of what is in a sense an energy coupling. The tapping-off of 3-phosphoglyceraldehyde to a lower energy state employs a process of negative $\Delta G^{0\prime}$ to drive reaction (4) which has a positive $\Delta G^{0\prime}$. The process of negative $\Delta G^{0\prime}$ is the series of reactions (Chs. 10 and 11) which accomplish the stepwise oxidation of the triose to three molecules of carbon dioxide.

Coupling of oxidative and reductive reactions

We have seen that many energy-yielding reactions are oxidative in character. Conversely, many energy-requiring reactions involve reduction.

For example, in fat synthesis (p. 132) a derivative of crotonic acid is converted to a derivative of butyric acid.

$$CH_3.CH=CH.COOR + 2H \rightleftharpoons CH_3.CH_2.CH_2.COOR \quad (13)\dagger$$

$$\Delta G^{0\prime} = +27\,700 \text{ calories/mole.}$$

It would presumably be possible to devise ways in which the hydrolysis of approximately four ATP molecules ($\Delta G^{0\prime} = 29\,000$) could be used to bring about the net reaction from left to right. This does not in fact occur. Instead the reducing power of reaction (8) proceeding from left to right is coupled directly to reactions of this type. This is done as an alternative to the trapping of the ΔG of reaction (8) in the form of ATP by means of oxidative phosphorylation.

Some of the processes of the type given in equation (8) lead to oxidative phosphorylation. The first step of many of these is as follows,

$$AH_2 + X \text{ (oxidized)} \rightleftharpoons A + X \text{ (reduced)} \quad (14)$$

when X is one or other of the closely related molecules, nicotin-amide adenine dinucleotide (NAD) and nicotinamide adenine dinucleotide phosphate (NADP). The structure of NAD is given in Table 3.2. That of NADP is the same with the addition of a phosphate group at the position indicated by *.

The reaction

$$NAD \text{ or } NADP \text{ (reduced)} + \tfrac{1}{2}O_2 \rightleftharpoons NAD \text{ or } NADP \text{ (oxidized)} \quad (15)$$

has a $\Delta G^{0\prime}$ of about $-40\,000$ calories/mole and represents the completion of the oxidation of AH_2 by molecular oxygen. The reduced forms of NAD and NADP are thus rather special examples of high energy compounds. We shall see in Ch. 8 that NAD (reduced) usually enters the remainder of the oxidative phosphorylation process and the energy of equation (15) is

† The nature of and need for the R group is explained fully in Ch. 16.

employed for the synthesis of three molecules of ATP (there is clearly more than enough energy available for this). NADP (reduced) does not usually enter oxidative phosphorylation but is used by coupling the equation

$$\text{NADP (reduced)} \rightleftharpoons \text{NADP (oxidized)} + 2\text{H} \qquad (16)$$

to equations like (13). We thus have

crotonyl − R + NADP (reduced) $\rightleftharpoons$ butyryl − R + NADP (oxidized).

$\Delta G^{0'}$ is $-40000 + 27000 = -13000$ calories/mole and so a net synthesis of butyryl − R is now easily achieved.

This is a typical example of the general rule that reduced NADP in distinction to reduced NAD is usually used for coupling to synthetic reactions (pp. 111, 120, 124 and 133). It is important to note that reduced NADP is the other immediate product, beside ATP, of the fall of the electron in photosynthesis. The photosynthetic process is needed for the synthesis of carbohydrate, and reduced NADP and ATP are used directly in this synthesis (see Ch. 15).

Summary

The change in the Gibbs free energy, ΔG, is a valuable measure of the availability of, or need for, energy in a reaction. ΔG is related to the equilibrium constant and this makes it possible to use the energy term to calculate the direction of net reaction.

Living matter operates energetically disfavoured reactions by coupling them with energetically favoured ones. The fall of a light-energized electron back to its ground state and the fall of electrons to molecular oxygen are the principal among the fundamental energy-yielding processes. These processes produce 'high energy' compounds. The high energy compounds are used to couple the fundamental energy sources to the energy-requiring processes of the cell.

8 The oxidative pathway

We have seen in Ch. 7 that oxidative processes are prominent among the energy sources available to the cell. Oxidations of organic compounds in which molecular oxygen is used as the final electron acceptor will yield as much as 50000 calories/atom of oxygen. In biological oxidations this yield of free energy is coupled to chemical synthesis and is not primarily expressed in the production of heat. The principal biochemical medium of energy exchange is ATP (p. 69). The synthesis of this molecule from ADP requires 7400 calories/mole and thus several molecules could result from one, efficiently coupled oxidation. A multi-stage process known as 'oxidative phosphorylation' exists for this purpose. To carry out the process in stages has the advantage that the energy may be taken off in steps nearer in size to those required for the production of a single molecule of ATP. Three molecules of ATP are produced by the full process, an energy yield of nearly 50%. This compares favourably with the efficiency of the reciprocating steam engine (about 7%) and with the overall efficiency of the steam turbine-dynamo couple (about 30%).

Other modes of biological oxidation do exist and some involve energy coupling. However, oxidative phosphorylation is of unique importance. It is a common adjunct to the majority of degradative reaction pathways described in this book and the principal source of the ATP required by the cell.

In the cells of higher organisms oxidative phosphorylation takes place, as was mentioned on p. 69, in an organized structure within the cell known as the mitochondrion (Frontispiece). There are many similarities between the oxidative production and the photo-synthetic production of ATP. The latter takes place in an ana-logous entity in photosynthetic cells, the chloroplast. In both cases the principal catalysts are protein – prosthetic group complexes (p. 28). Lipids (p. 56) also are present. It is possible that lipids have an active role in the phosphorylation process; they are certainly extremely important in establishing and controlling the critical three-dimensional inter-relationships between the other components.

We will now describe, in a simplified form, the stages of oxidative phosphorylation. Almost certainly much more awaits discovery, but the general outline of our understanding of the process is unlikely to change to a significant extent.

The hydrogen and electron transport pathway

The oxidation of an organic compound may be written

$$AH_2 + \tfrac{1}{2}O_2 \rightleftharpoons A + H_2O. \tag{1}$$

For the majority of compounds the first stage is

$$AH_2 + NAD \text{ (oxidized)} \rightleftharpoons A + NAD \text{ (reduced)}. \tag{2}$$

This will be an enzyme-catalysed reaction. The enzyme may either be absolutely specific for A or catalyse the dehydrogenation of a range of similar compounds.

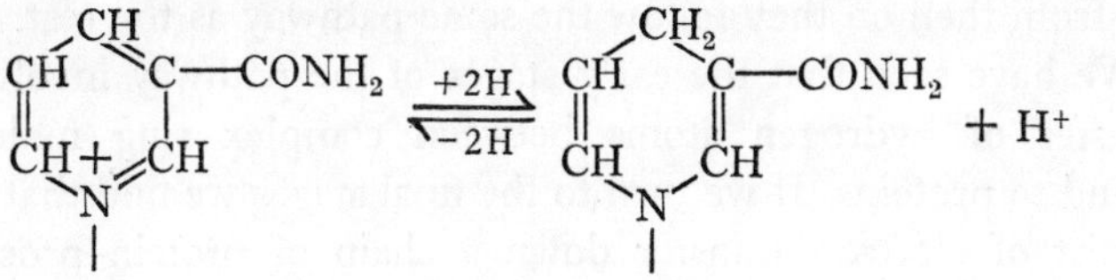

Fig. 8.1. Oxidation and reduction in the nicotinamide ring.

NAD (p. 72, Fig. 8.1) is usually bound (non-covalently) to the enzyme catalysing reaction (2). The strength of the binding is more characteristic of a cofactor than a prosthetic group. The part of the molecule which is concerned in reduction–oxidation is the nicotinamide ring (Fig. 8.1). (It should be possible to see from this figure why the oxidized form of NAD is sometimes written as NAD^+ and the reduced form NADH, with a similar notation for NADP. We prefer the notation $NADH_2$ for the reduced form and NAD for the oxidized form.)

It is beyond the capacity of human metabolism to synthesize nicotinamide. The vital place of NAD and NADP in metabolism explains why nicotinamide is a vitamin (see also p. 28).

If coupling to a reducing reaction (p. 72) is not to occur, the next stage is the reduction by $NADH_2$ of another ring system (Fig. 8.5). The ring is found in nature as part of two molecules, flavin mononucleotide (FMN) and flavin adenine dinucleotide (FAD) (see Table 3.2). FMN and FAD are attached to proteins by non-covalent forces. The attachment is strong enough for them to be termed prosthetic groups. The FMN and FAD–protein complexes are known as flavoproteins.

Like nicotinamide riboflavin cannot be synthesized by man and is a vitamin (B_2).

The flavoprotein involved in the reaction

$$NADH_2 + \text{flavoprotein (oxidized)} \rightleftharpoons NAD + \text{flavoprotein (reduced)} \quad (3)$$

can be thought of as an enzyme catalysing the dehydrogenation of $NADH_2$. Some flavoproteins are, in fact, dehydrogenase enzymes in the more usual sense – for example, the desaturation of aliphatic carbon

$$R_1 . CH_2 . CH_2 . R_2 \rightleftharpoons R_1 . CH = CH . R_2$$

(see pp. 99 and 104). Such dehydrogenations therefore bypass NAD but from then on they follow the same pathway as the rest.

We have seen that the early stages of the pathway involve the transfer of hydrogen atoms between complex ring molecules bound to proteins. If we turn to the final stages we find that these consist of electron transfer down a chain of protein–prosthetic group complexes (the cytochromes, see below), ultimately ending in oxygen.

The cytochromes rely on ferro-porphyrin rings similar to those used by haemoglobin. The three main families of cytochromes (cytochromes *a*, *b* and *c*) differ from one another by virtue of their selection of R groups at the corners of the ring (Fig. 8.2). Individual members of a family differ from one another in their amino-acid sequence and not necessarily in the ring substitution.

Cytochrome *c* is the best studied of the cytochromes and is probably typical. In distinction to haemoglobin (p. 37) the ring is covalently as well as non-covalently bound to the protein. The sixth co-ordination position of the iron, which in haemoglobin was free to receive oxygen, is permanently occupied by an amino-acid side chain. Rather than being exposed on the surface of the protein, the ring is buried in a crevice and only one edge lies on the surface

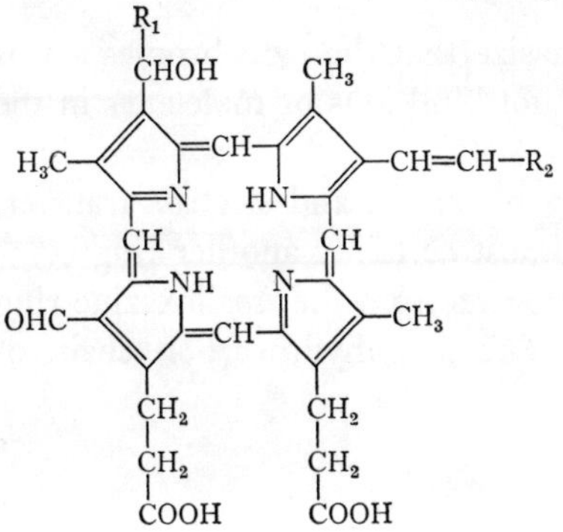

Cytochrome *a*

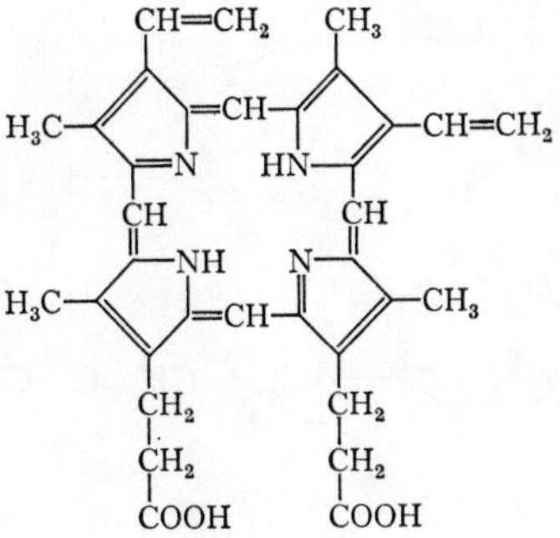

Cytochrome *b*

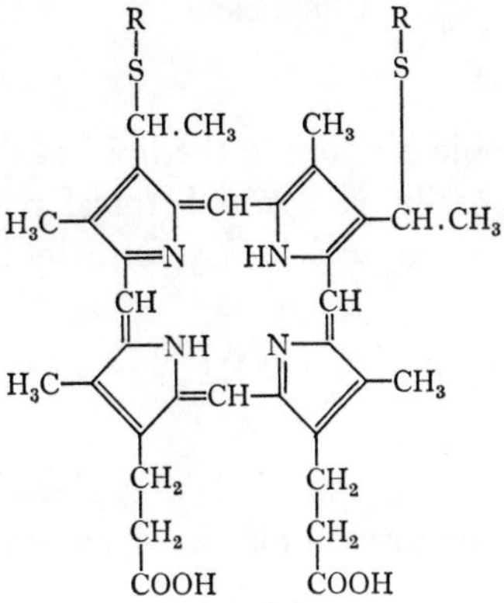

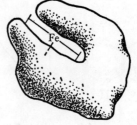

Fig. 8.3. Cytochrome *c*.

Cytochrome *c*

Fig. 8.2. Substitutions in the
porphyrin rings of the cytochromes.

(Fig. 8.3). This serves to emphasize that the cytochromes act as conductors of electrons and do not bind ions or molecules in the way that haemoglobin does.

How the two systems, hydrogen transfer and electron transfer, are linked is not quite clear. Almost certainly another ring compound, ubiquinone, accepts hydrogens from the *iso*alloxazine ring of the flavoprotein (Fig. 8.4). The long hydrocarbon chain of

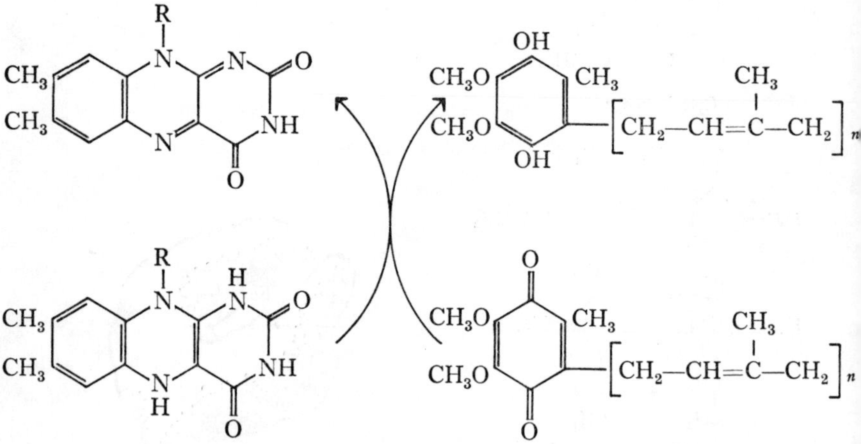

(For the nature of R see Table 3.2)

*iso*alloxazine part of FMN or FAD   Ubiquinone

Fig. 8.4

ubiquinone would favour hydrophobic binding to the lipid of the mitochondrion. This is the last easily recognized transfer of hydrogen atoms, thereafter a step occurs which may formally be written as:

$$2H \text{ (from the hydrogen transfer pathway)} \rightleftharpoons 2H^+ + 2e^- \qquad (4)$$

and electron transfer commences.

We have already seen that the whole process is taking place in a highly unusual environment, a lipid–protein matrix. Therefore, while equation (4) and equation (5) below are useful for a general understanding of what takes place, we need not necessarily assume that the protons and electrons ever exist as identifiable separate

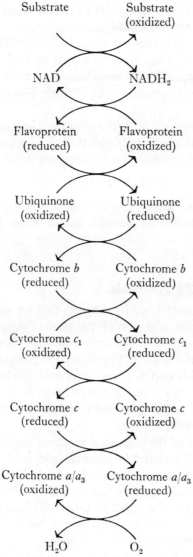

Fig. 8.5. The hydrogen and electron transport pathway.

entities. Also, the succession of reactions is so fast that once again (see p. 36) we are led to assume that there is no free diffusion of intermediates from one protein to another. The active part of the carriers must be almost in contact and the product of one reaction handed directly on to the catalyst of the next.

How reaction (5) is brought about is far from clear except that other, metal-containing, proteins appear to be involved. Thereafter, matters appear to be straightforward with the transfer of electrons from one cytochrome to another. The order is given in Fig. 8.5.

Finally cytochrome a or a_3 (the separate existence of cytochrome a and a_3 has been questioned) with the help of a Cu^{2+} ion catalyses the final reaction in which the electrons are donated to molecular oxygen

$$2e^- + 2H^+ + \tfrac{1}{2}O_2 \rightleftharpoons H_2O \tag{5}$$

and the oxidative pathway is at an end. (Cyanide is an inhibitor of the terminal cytochrome and this fact accounts for its extreme toxicity.)

The sites of ATP synthesis

Those oxidations catalysed by enzymes that are also flavoproteins yield only two molecules of ATP per atom of oxygen used. Those involving NAD-linked dehydrogenases yield three as does the oxidation of $NADH_2$ itself. Therefore, one phosphorylation must be associated with step 3. A variety of evidence has been used to locate the other two sites. One appears to be between the ubiquinone/cytochrome b region and cytochrome c_1, the other on the oxygen side of cytochrome c.

The energetics of the individual reactions are difficult to determine. The normal electrode potential and $\Delta G^{0\prime}$ determinations can be carried out for the molecules involved when in free solution: as has been pointed out more than once these are not the circumstances obtaining in the cell. The values obtained in free solution are likely to be greatly modified in the 'solid state' but some estimate can be made of the true situation. Thus, while somewhat speculative, Fig. 8.6 shows that some steps do have the energy span necessary for the phosphorylation, while others certainly do not. On the whole there is good agreement between the energetic calculations and all other experimental data on the location of the phosphorylation sites.

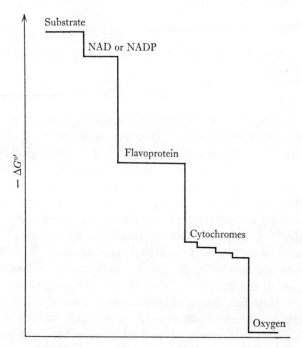

Fig. 8.6. Redox potentials in hydrogen and electron transport (idealized).

The mechanism of the phosphorylation

The transfer of hydrogen, and more particularly electrons, seems at first sight to have little to do with the reaction:

$$ADP + phosphate \rightleftharpoons ATP.$$

In fact all chemical reactions are electron processes and we should not find the connection too hard to imagine. Imagine is the word, unfortunately, because in spite of much experimental work little has emerged that is definite. The reader is referred to more advanced books for a discussion of the possible mechanisms which have been suggested. However, the important thing is that electrons are flowing from a higher state of available free energy to a lower and that the energy delivered by this flow is coupled to the synthesis of ATP.

9 Introduction to intermediary metabolism

We have now set the scene for a discussion of some aspects of intermediary metabolism. This name is given to that set of reactions by which the small molecules found in biological systems are degraded, synthesized, interconverted and otherwise chemically transformed. A striking characteristic of these transformations is that they occur in very small, discrete steps, each of which is catalysed by a separate enzyme. Thus intermediary metabolism consists of a large number of reactions, by means of which molecules are gradually modified and shaped. The various reactions are generally thought of as comprising 'pathways', along which the compounds flow while undergoing these gradual changes. However, the pathways constantly converge and diverge, so that an intermediate formed as a result of one reaction may often have a choice of two or more subsequent paths to follow.

One reason why metabolic transformations involve so many steps will become clear if we consider the mode of action of enzymes. If metabolism is to be at all vigorous, it is necessary for enzymes to be as active as possible. In order to increase the rate of a biochemical reaction the enzyme must bind to the substrate and to the reaction intermediate (p. 45). For the increase in rate to be appreciable, the substrate and subsequently the reaction intermediate must be very strongly bound through a number of firm, specific contacts with the enzyme. It is remarkable that a linear polymer of amino acids can do so well in providing a binding surface even for normal enzyme-catalysed reactions, in which only a few atoms are involved: to provide a surface that could organize (for example) the simultaneous oxidation of all six carbon atoms of glucose would be considerably more difficult.

Thus as a result of the immense catalytic power that enzymes possess, the chemical change that any one enzyme reaction brings about is correspondingly slight. It seems to be for this reason that

[82]

the pathways of intermediary metabolism involve such gradual modification of the reacting molecules.

The fact that any one enzyme does such a specific, small-scale job necessitates the presence in organisms of a large number of different enzymes. One might think that, even despite the points that we made above, it would be better for organisms not to have to make so many enzymes. For example, we shall see in Ch. 10 that the breakdown of glucose to lactic acid requires eleven separate enzymes, each of which must be made by the specific protein-synthesizing machinery of the cell (Ch. 20). Would it not be more economical to have a single enzyme that achieves the production of lactic acid from glucose, even if that enzyme had a much more complex structure and, perhaps, less catalytic power than the enzymes that actually exist?

The answer is that this division of metabolic reactions into small steps is admirably suited to fulfilling two separate functions undertaken by intermediary metabolism. These are to provide energy in the form of ATP and to produce the building blocks from which macromolecules are synthesized. In the degradation of glucose to lactic acid several intermediates are formed which are used in other pathways – for example glucose-6-phosphate which can be converted to glycogen (Ch. 15), 3-phosphoglyceraldehyde which can be converted to glycerol and thus used in the synthesis of fats (Ch. 16), and pyruvic acid which (among many other uses) can be converted to alanine and used in the synthesis of proteins (Ch. 20). A similar multiple use is made of every pathway in intermediary metabolism. If glucose were degraded to lactic acid in one step the essential intermediates would have to be formed by other routes, each of which would require its own separate enzyme or enzymes. So the advantage that would seem to result if enzymes were designed to achieve much more radical transformations would in fact be no advantage at all.

On the contrary, there is yet another advantage in splitting up metabolic pathways into small steps. The economy of the cell demands that the processes of intermediary metabolism be carefully controlled, so that the ATP and intermediates that they provide should be available in the right quantities at the right time. Now by dividing metabolic pathways into a large number of steps the cell can effect a very fine control by altering the rate of just a single reaction. For example, we shall see that the degradation of glucose to lactic acid involves a reaction catalysed by the enzyme phosphofructokinase (p. 90). This enzyme, however, is needed equally for the degradation of lactose via galactose and of glycogen, in each case to lactic acid. So by blocking the activity of phosphofructokinase the cell can block the degradation of all carbohydrate (p. 184) without at the same time affecting any other metabolic process.

Although other general points about intermediary metabolism could be made, we have said enough to explain some of the reasons for what might seem a bewildering complexity of metabolic reactions. In the chapters that go into intermediary metabolism in some detail we have been obliged to separate, for the purposes of exposition, the two functions of the reactions that we describe – the formation of ATP and the production of useful intermediates for biosynthesis. But it is important not to lose sight of their inter-relationship, and in order to give an overview of the subject we shall now summarize the contents of Chs. 10–17, so that each pathway can be seen in its context.

If we first consider the chief energy-yielding pathways, we can draw a chart (Fig. 9.1) that gives the outline of the processes. Polysaccharides are the principal storage molecules (Ch. 6); and these are depolymerized to hexoses (sugars that contain six carbon atoms). Hexoses are broken down *via* C_3 sugars (sugars that contain three carbon atoms) to C_3 acids, as we shall see in Ch. 10. C_3 acids are oxidized to give acetyl (C_2) fragments, and these join with a C_4 acid to give a C_6 acid, which is then successively oxidized *via* a C_5 acid to yield a C_4 acid again which can once more accept an acetyl fragment (Ch. 11); this sequence of reactions thus completely oxidizes the C_3 acids, and hence the hexoses. The other important storage molecules are fats, which are composed of glycerol and fatty acids. Glycerol is converted to a C_3 sugar; fatty acids are oxidized to acetyl fragments which are degraded as before (Ch. 12).

During the course of these degradations, pairs of hydrogen atoms are removed from the intermediates at several stages in the pathways. These pairs of hydrogen atoms are generally passed to NAD and thus to the oxidative phosphorylation pathway (Ch. 8), and ATP results from their oxidation. The one pathway that reduces NADP rather than NAD is the oxidation of a hexose to a pentose (C_5 sugar) (Ch. 13).

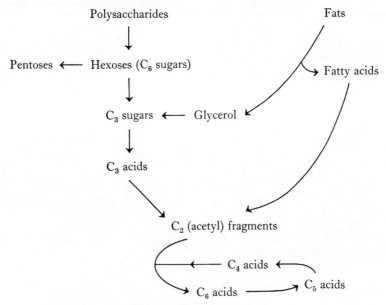

Fig. 9.1. Sketch-map of metabolic reactions to show the degradation of carbohydrates and fats.

By contrast, photosynthetic organisms derive both ATP and $NADPH_2$ by conversion of the energy of sunlight (Ch. 14), and they use these compounds to fix carbon dioxide into a C_3 sugar (Ch. 15). C_3 sugars can be converted into hexoses both in photosynthetic and non-photosynthetic organisms; but (as indicated in Fig. 9.2) the pathway of synthesis of hexoses is in part different from the path of degradation, and the same is true for the synthesis of polysaccharides from hexoses (Ch. 15). Hexoses can also be formed from C_3 acids that derive from C_4 acids (Ch. 15). In addition, as Fig. 9.2 shows, fatty acids can be formed from acetyl fragments; but the pathway of synthesis is different from the degradative pathway, and this is true also of the formation of the fats themselves. Equally the conversion of pentoses to hexoses (Ch. 13) is not just the reverse of the conversion of hexoses to pentoses.

We can now see how the intermediates that are formed in these pathways are used in the synthesis of the building blocks for macromolecules – the amino acids that are constituents of proteins (p. 5) and the purine and pyrimidine nucleotides that are

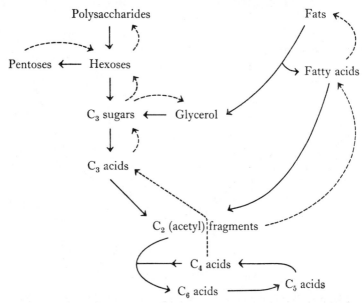

Fig. 9.2. Sketch-map of metabolic reactions to show the degradation and synthesis of carbohydrates and fats.

constituents of nucleic acids (p. 6). Fig. 9.3, which is an expansion of the left-hand side of Fig. 9.1 and Fig. 9.2, shows some of these synthetic pathways. Amino acids can be made from several of the compounds that are produced from the breakdown of carbohydrates, such as C_3 acids and, more important, C_4 and C_5 acids (Ch. 17). The C_4 amino acid aspartic acid gives rise to other amino acids (Ch. 17) and also to pyrimidines (Ch. 19); the C_5 amino acid glutamic acid gives rise to still other amino acids (Ch. 17). The skeleton of purines (unlike that of pyrimidines) is formed piecemeal from a number of precursors (Ch. 19). The ribose phosphate and deoxyribose phosphate that are constituents of nucleic acids are formed from the pentose produced by oxidation of hexose (Ch. 13).

The next seven chapters describe in much more detail the pathways that we have mentioned here. You will find that these chapters are extensively cross-referenced: it is our intention to stress the inter-relationships of the metabolic pathways that we

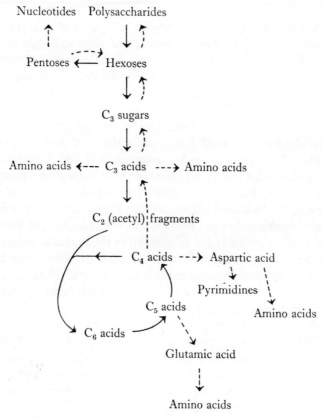

Fig. 9.3. Sketch-map of metabolic reactions to show how the synthesis of amino acids and nucleotides is related to intermediates in the degradation of carbohydrates.

discuss, and thus to show how intermediary metabolism combines the two functions that we have mentioned – providing energy and supplying the intermediates needed for the synthesis of macromolecules.

10 Synthesis of ATP – glycolysis

Glucose is the key sugar in carbohydrate metabolism. We have previously seen (Ch. 6) that starch, the polysaccharide storage material of plants, and glycogen, the polysaccharide storage material of animals, both consist of glucose residues. Moreover, other sugars can be converted to glucose by simple series of reactions. Again, carbohydrate is supplied in the form of glucose to most cells of the animal body by the blood. Thus in discussing the breakdown of carbohydrates we shall first consider the breakdown of glucose and then show how some other compounds fit into the scheme.

Organisms that live under anaerobic conditions (that is to say, in the absence of oxygen) can break down glucose by *fermentation* reactions. In the course of fermentation the molecule is split without undergoing either net oxidation or net reduction, so that the number of hydrogen and oxygen atoms in the products is the same as in the starting material. At first sight fermentations appear to be very varied, because different organisms accumulate different products of fermentation – lactic acid, ethanol, butanol, acetone, propionic acid, etc., etc. But this apparent diversity conceals the fact that most of the reactions in the pathway are actually the same for different organisms; it is only at the very last stages that the pathways in different species diverge. So we are justified in looking, to begin with, at just one fermentation as an example of the anaerobic breakdown of glucose. The most convenient one to examine is the formation of lactic acid.

Although the fermentation of glucose to produce lactic acid is an anaerobic process, that does not mean that it is of no importance to organisms that make use of molecular oxygen to oxidize carbohydrates. On the contrary, aerobic organisms have in fact adopted the reactions of the fermentation pathway in order to break down glucose into smaller fragments that they can then oxidize. In aerobic organisms like ourselves, anaerobic conditions sometimes prevail temporarily in the muscles, and lactic acid appears in quite large quantities in actively working muscle. Even in tissues that never become anaerobic much of the metabolism of glucose still involves borrowing the reactions of the fermentation pathway. These reactions split the glucose molecule in halves, and the reactions of the aerobic pathway (see next chapter) then oxidize the fragments thus formed.

Glycolysis

The anaerobic breakdown of glucose is often called *glycolysis*, and it proceeds via a pathway often called the Embden–Meyerhof pathway. Although the detailed reactions of the Embden–Meyerhof pathway are quite complicated, the net result is extremely simple. 'Glycolysis' is the splitting of glucose, and the sum total of the reactions is just this:

$$C_6H_{12}O_6 \longrightarrow 2C_3H_6O_3,$$
$$\text{glucose} \longrightarrow 2 \text{ lactic acid.}$$

It is evident that this reaction involves no net oxidation or reduction, and is therefore a true fermentation.

Let us now look at the details of this process. Before glucose can undergo any metabolic change, it must first be phosphorylated by an enzyme called hexokinase.

glucose   glucose-6-phosphate

The $\Delta G^{0\prime}$ of this reaction overwhelmingly favours the formation of glucose-6-phosphate. But this fact does not mean that the organism has no means of forming glucose from glucose-6-phosphate, since there is a phosphatase capable of hydrolysing glucose-6-phosphate. However, although hexokinase is widely distributed in living organisms and is found throughout the mammalian body, phosphatase in the mammal is largely confined to the liver and is absent, for example, from muscle.

What can we conclude from these facts? It seems that one of the important features of the phosphorylation of glucose is that glucose itself can diffuse through cell membranes but glucose-6-phosphate cannot. By phosphorylating glucose that has diffused in from the blood, a cell prevents it from leaking out; and thus a tissue like muscle can readily trap the glucose that it needs. A liver cell, on the other hand, has the function of supplying glucose to the blood. In order to do so it breaks down glycogen to form glucose-6-phosphate (see p. 95), and it must then cleave off phosphate by use of the phosphatase, to allow the non-phosphorylated glucose to leave the cell and enter the blood.

Glucose-6-phosphate is in equilibrium in the cell with fructose-6-phosphate, the equilibrium being catalysed by the enzyme hexose phosphate isomerase (see p. 64). Fructose-6-phosphate can now be phosphorylated once more at the expense of ATP. The reaction is catalysed by phosphofructokinase and yields fructose-1;6-diphosphate. As with the phosphorylation of glucose the $\Delta G^{0\prime}$ greatly favours the formation of the product, and in order to break down fructose-1;6-diphosphate to fructose-6-phosphate an enzyme (fructose diphosphatase) catalysing a different reaction is required.

glucose-6-phosphate   fructose-6-phosphate   (2)

fructose-6-phosphate   fructose-1;6-diphosphate   (3)

Fructose-1;6-diphosphate can be cleaved by the action of the enzyme aldolase to give two molecules of triose phosphate. One of these is dihydroxyacetone phosphate and the other 3-phosphoglyceraldehyde. These two molecules are themselves in equilibrium in the cell. The attainment of equilibrium is catalysed by the enzyme triose phosphate isomerase (see p. 65). As we have already discussed (p. 65) the next reaction in the glycolytic sequence uses only 3-phosphoglyceraldehyde but because of the equilibrium just mentioned dihydroxyacetone phosphate is converted into 3-phosphoglyceraldehyde as the latter is used up. Thus *both* of the triose phosphates derived from fructose-1;6-diphosphate are in fact further metabolized.

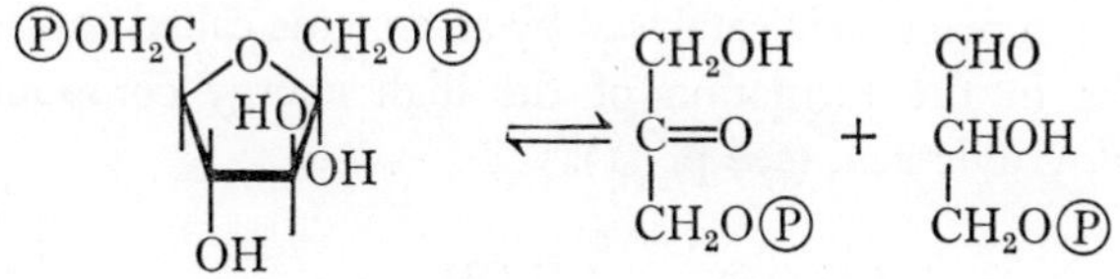

$$\underset{\substack{\text{fructose-1;6-}\\\text{diphosphate}}}{\text{(P)}OH_2C \underset{OH}{\overset{O}{\diagup}} CH_2O\text{(P)} \\ HO\diagdown OH} \rightleftharpoons \underset{\substack{\text{dihydroxyacetone}\\\text{phosphate}}}{\overset{CH_2OH}{\underset{CH_2O\text{(P)}}{C=O}}} + \underset{\substack{\text{3-phosphoglyceraldehyde}}}{\overset{CHO}{\underset{CH_2O\text{(P)}}{CHOH}}} \quad (4)$$

$$\underset{\text{dihydroxyacetone phosphate}}{\overset{CH_2OH}{\underset{CH_2O\text{(P)}}{C=O}}} \rightleftharpoons \underset{\text{3-phosphoglyceraldehyde}}{\overset{CHO}{\underset{CH_2O\text{(P)}}{CHOH}}} \quad (5)$$

We have now arrived at a most significant reaction of glycolysis, that in which 3-phosphoglyceraldehyde is oxidized to 1;3-diphosphoglyceric acid. The reaction is catalysed by the enzyme triose phosphate dehydrogenase. This reaction is of particular importance because it results in the formation of a high energy compound. This is an example of substrate-level phosphorylation (see p. 71).

$$\underset{\text{3-phosphoglyceraldehyde}}{\overset{CHO}{\underset{CH_2O\text{(P)}}{CHOH}}} + NAD + P_i \rightleftharpoons \underset{\text{1;3-diphosphoglyceric acid}}{\overset{COO\text{(P)}}{\underset{CH_2O\text{(P)}}{CHOH}}} + NADH_2 \quad (6)$$

$$
\begin{array}{ccc}
\begin{array}{l} \text{CH}_2\text{O}\,\textcircled{P} \\ | \\ \text{CHOH} \\ | \\ \text{COO}\textcircled{P} \end{array} + \text{ADP} \rightleftharpoons
\begin{array}{l} \text{CH}_2\text{O}\,\textcircled{P} \\ | \\ \text{CHOH} \\ | \\ \text{COOH} \end{array} + \text{ATP}
\end{array} \tag{7}
$$

At first sight the oxidation of 3-phosphoglyceraldehyde and the accompanying reduction of NAD appear to conflict with the principle that we mentioned earlier – that glycolysis does not involve net oxidation or reduction of the glucose molecule. But we shall see shortly that the hydrogen is stored only temporarily as $NADH_2$; it will soon be returned to the main pathway.

3-Phosphoglyceric acid is changed (by an enzyme called a mutase) to 2-phosphoglyceric acid, and this is now dehydrated. The dehydration reaction is catalysed by an enzyme called enolase, and it results in the formation of the high energy compound phospho-*enol*pyruvic acid (see p. 71).

$$
\begin{array}{cc}
\begin{array}{l} \text{CH}_2\text{O}\textcircled{P} \\ | \\ \text{CHOH} \\ | \\ \text{COOH} \end{array} \rightleftharpoons
\begin{array}{l} \text{CH}_2\text{OH} \\ | \\ \text{CHO}\textcircled{P} \\ | \\ \text{COOH} \end{array}
\end{array} \tag{8}
$$

3-phosphoglyceric acid   2-phosphoglyceric acid

$$
\begin{array}{cc}
\begin{array}{l} \text{CH}_2\text{OH} \\ | \\ \text{CHO}\textcircled{P} \\ | \\ \text{COOH} \end{array} \rightleftharpoons
\begin{array}{l} \text{CH}_2 \\ \| \\ \text{CO}\textcircled{P} + \text{H}_2\text{O} \\ | \\ \text{COOH} \end{array}
\end{array} \tag{9}
$$

2-phosphoglyceric acid   phospho-*enol*pyruvic acid

As we explained on p. 66 the free energy of hydrolysis of phospho-*enol*pyruvic acid is very high, and the enzyme pyruvate kinase can readily catalyse the synthesis of ATP. (We mentioned that the free energy of hydrolysis of phospho-*enol*pyruvic acid is so exceptionally high that the reaction is not reversible to any great extent.) The removal of the phosphate group from phospho-*enol*pyruvic acid leaves *enol*pyruvic acid, which tautomerizes without the intervention of an enzyme to *keto*pyruvic acid (commonly called pyruvic acid).

$$
\begin{array}{ccc}
\text{CH}_2 & & \text{CH}_2 \\
\parallel & & \parallel \\
\text{CO}\textcircled{P} + \text{ADP} \rightleftharpoons & \text{COOH} + \text{ATP} & \quad (10) \\
\mid & & \mid \\
\text{COOH} & & \text{COOH}
\end{array}
$$

phospho-*enol*pyruvic acid   *enol*pyruvic acid

$$
\begin{array}{l}
\text{CH}_3 \\
\mid \\
\text{CO} \qquad \textit{keto}\text{pyruvic acid} \\
\mid \\
\text{COOH}
\end{array}
$$

The glycolytic pathway is nearly complete – nearly, but not quite. Reaction (6) of the pathway reduced NAD to $NADH_2$; and since the concentration of NAD in cells is very low it is evident that glycolysis could not continue for any length of time without regeneration of NAD to act once again as a hydrogen carrier in reaction (6). Under aerobic conditions it would be possible to reoxidize $NADH_2$ by the respiratory chain (p. 76). But glycolysis is an *anaerobic* pathway, and is the principal means of ATP synthesis for many organisms that do not use molecular oxygen for oxidation. How then can NAD be regenerated from $NADH_2$ anaerobically?

The answer is that $NADH_2$ is reoxidized by being used to reduce the product of the glycolytic pathway itself. Pyruvic acid was formed in reaction (10), and the enzyme lactate dehydrogenase readily catalyses the reduction of pyruvic acid to lactic acid.

$$
\underset{\text{pyruvic acid}}{\text{CH}_3.\text{CO}.\text{COOH}} + NADH_2 \rightleftharpoons \underset{\text{lactic acid}}{\text{CH}_3.\text{CHOH}.\text{COOH}} + NAD. \quad (11)
$$

The NAD can now be used again in reaction (6), and in this way, by recycling NAD, a cell can break down large quantities of glucose and accumulate lactic acid.

Some organisms accumulate not lactic acid but other products of fermentation. This fact need not surprise us if we remember that the function of the last reaction we have mentioned (11) is to dispose of excess hydrogen and to reoxidize $NADH_2$. Thus any compound that can accept hydrogen from $NADH_2$ can act in place of the pyruvic acid in reaction (11). For example yeast decarboxylates the pyruvic acid first and then uses the product of decarboxylation to reoxidize $NADH_2$.

$$\begin{array}{c} CH_3 \\ | \\ CO \\ | \\ COOH \end{array} \overset{\rightharpoonup}{\leftharpoondown} CO_2 + \begin{array}{c} CH_3 \\ | \\ CHO \end{array} \xrightarrow{\underset{NADH_2\ \ NAD}{\frown}} \begin{array}{c} CH_3 \\ | \\ CH_2OH \end{array} \quad (12)$$

pyruvic acid   acetaldehyde   ethanol

Other organisms carry out more complex reactions which produce a variety of reduced products, but the biological function is the same in each case.

Many organisms that live anaerobically derive all of their ATP from fermentations of this kind. We can easily calculate the net yield of ATP from glycolysis. One molecule of ATP is used up in each of the reactions (1) and (3). Reaction (4) splits the molecule in halves and since one ATP is gained in each of the reactions (7) and (10) for each half molecule of glucose, the gain from these two reactions is four ATP. Thus the *net* gain of ATP from the fermentation of glucose to lactic acid (or other fermentation products) is two molecules of ATP.

Other compounds and the Embden–Meyerhof pathway

There are several compounds other than glucose that are degraded by the Embden–Meyerhof pathway, although in general two or three special reactions are needed to bring the compound into the pathway. *Fructose*, for example, can be phosphorylated directly to fructose-6-phosphate and enter the pathway at reaction (3). *Galactose* is phosphorylated and eventually converted to glucose-6-phosphate. The glycolytic pathway can also be used for the break-down of *glycerol* in the following way, and the dihydroxyacetone phosphate thus formed can enter at reaction (5).

$$\begin{array}{c} CH_2OH \\ | \\ CHOH \\ | \\ CH_2OH \end{array} \xrightarrow{\underset{ATP\ \ ADP}{\frown}} \begin{array}{c} CH_2O\textcircled{P} \\ | \\ CHOH \\ | \\ CH_2OH \end{array} \xrightarrow{\underset{NAD\ \ NADH_2}{\frown}} \begin{array}{c} CH_2O\textcircled{P} \\ | \\ CO \\ | \\ CH_2OH \end{array}$$

glycerol   α-glycerophosphate   dihydroxyacetone
                    phosphate

In animals one very important use of the Embden–Meyerhof pathway is to provide energy by the breakdown of glycogen (see p. 56). (In plants starch replaces glycogen.) In animal muscles, particularly, an abrupt demand for ATP is met by glycolysing glycogen without going through glucose as an intermediate. The first step is depolymerization of glycogen by reaction with inorganic phosphate, catalysed by the enzyme phosphorylase.

$$(\text{glucose})_n + P_i \rightleftharpoons (\text{glucose})_{n-1} + \text{glucose-1-phosphate}.$$

This reaction does not use ATP, so that there is no need for a kinase to bring the glucose residues into the metabolic pathway. (By contrast, phosphorylation of the sugars such as glucose, fructose and galactose, and also of glycerol, *does* require a kinase and uses ATP.) A mutase now catalyses the transfer of the phosphate group from the 1 to the 6 carbon atom of glucose.

glucose-1-phosphate   glucose-6-phosphate

The glucose-6-phosphate can enter the Embden–Meyerhof pathway at reaction (2). Alternatively, in liver the glucose-6-phosphate thus formed from glycogen can be split to glucose (see p. 89).

11 Synthesis of ATP – the Krebs cycle

Pyruvic acid is the penultimate compound of the glycolytic pathway, and we have seen that under anaerobic conditions the $NADH_2$ produced in reaction (6) of this pathway is reoxidized to NAD by reducing pyruvic acid. In aerobic conditions, on the other hand, $NADH_2$ can be reoxidized to NAD via the respiratory chain (see p. 76), so that in this case there is no need for pyruvic acid to act as an acceptor of hydrogen. Instead, pyruvic acid is broken down by oxidation.

The first stage in the breakdown of pyruvic acid is an oxidative decarboxylation. This reaction proceeds in several steps, catalysed by the enzyme complex pyruvate dehydrogenase. In the first step an acetyl group is removed from pyruvic acid, leaving carbon dioxide, and then this acetyl group is passed from one carrier to another until it reaches the carrier coenzyme A.

The first of these molecules that carry the acetyl group is thiamine pyrophosphate. Its structure is given on p. 28, but we shall here write it as TPP. The next carrier is lipoic acid which we shall write

where $R=(CH_2)_4COOH$.

Coenzyme A (p. 28) we shall write CoASH. Thus, as far as the formation of acetyl coenzyme A from pyruvic acid is concerned we can write the reaction sequence as follows.

$$CH_3.CO.COOH + TPP \rightleftharpoons CH_3.CHO.TPP + CO_2$$

$$\begin{array}{c}
\underset{\underset{H}{|}}{S} \quad \underset{\underset{CO.CH_3}{|}}{S}\text{\textasciicircum}^R + CoASH \rightleftharpoons \underset{\underset{H}{|}}{S} \quad \underset{\underset{H}{|}}{S}\text{\textasciicircum}^R + CH_3.CO.S.CoA
\end{array}$$

We now have a difficulty similar to that which we saw in the accumulation of $NADH_2$ during glycolysis (p. 93). One result of the oxidative decarboxylation of pyruvic acid is that lipoic acid becomes reduced. Since lipoic acid is a coenzyme and is present only in small quantities in the cell, it must be reoxidized if it is to function in the breakdown of further molecules of pyruvic acid. In fact reduced lipoic acid can be reoxidized by NAD in a reaction catalysed by lipoate dehydrogenase.

$$\begin{array}{c}
\underset{\underset{H}{|}}{S} \quad \underset{\underset{H}{|}}{S}\text{\textasciicircum}^R + NAD \rightleftharpoons S\text{---}S\text{\textasciicircum}^R + NADH_2
\end{array}$$

Since all of these reactions are proceeding in the presence of oxygen $NADH_2$ is reoxidized by the respiratory chain. Thus the sum of the four reactions so far is

$$CH_3.CO.COOH + CoASH + NAD \rightleftharpoons$$
$$CH_3.CO.S.CoA + CO_2 + NADH_2$$

pyruvic acid   acetyl coenzyme A   (1)

Acetyl coenzyme A, at which we have now arrived, is one of the central compounds of intermediary metabolism. We have just seen how it is produced by oxidation of pyruvic acid, and hence it is an intermediate in the breakdown of carbohydrates and of glycerol. We shall see later (Ch. 12) that it is the breakdown product of fatty acids, and also (Ch. 17) of some amino acids. Again, acetyl coenzyme A is the starting point of synthesis for fatty acids (see Ch. 16) and for steroids and carotenoids.

We can now examine the fate of the acetyl group of acetyl coenzyme A. The oxidation of this acetyl group proceeds through a sequence of reactions that is frequently called the Krebs cycle, citric acid cycle or tricarboxylic acid cycle, which we outlined at the bottom of Figs. 9.1 and 9.2.

In the first reaction, acetyl coenzyme A reacts with a compound called oxaloacetic acid. This acid is, as we shall see shortly, formed in the *last* reaction of the Krebs cycle. Thus it acts, in a sense, as a carrier of the acetyl group through the reactions of the cycle and is left over at the end to accept another acetyl group.

$$
\text{CH}_3.\text{CO}.\text{S}.\text{CoA} + \underset{\overset{|}{\text{CH}_2.\text{COOH}}}{\text{CO}.\text{COOH}} + \text{H}_2\text{O} \rightleftharpoons \underset{\overset{|}{\text{CH}_2.\text{COOH}}}{\overset{\text{CH}_2.\text{COOH}}{\overset{|}{\text{HOC}}.\text{COOH}}} + \text{CoASH} \quad (2)
$$

acetyl coenzyme A   oxaloacetic acid   citric acid

This reaction is catalysed by citrate synthetase.

The citric acid is now isomerized to form isocitric acid. The enzyme responsible for the isomerization is called aconitase.

$$
\begin{array}{c}
\text{CH}_2.\text{COOH} \\
| \\
\text{HOC}.\text{COOH} \\
| \\
\text{CH}_2.\text{COOH} \\
\text{citric acid}
\end{array}
\rightleftharpoons
\begin{array}{c}
\text{CHOH}.\text{COOH} \\
| \\
\text{CH}.\text{COOH} \\
| \\
\text{CH}_2.\text{COOH} \\
\text{isocitric acid}
\end{array}
\qquad (3)
$$

Isocitric acid is dehydrogenated by NAD in a reaction catalysed by isocitrate dehydrogenase. The resulting oxalosuccinic acid is unstable and would decarboxylate spontaneously without an enzyme, but in fact the enzyme responsible for its formation also catalyses its decarboxylation.

$$
\begin{array}{c}
\text{CHOH}.\text{COOH} \\
| \\
\text{CH}.\text{COOH} \\
| \\
\text{CH}_2.\text{COOH} \\
\text{isocitric acid}
\end{array}
\xrightarrow{\overset{\text{NAD} \quad \text{NADH}_2}{\quad}}
\begin{array}{c}
\text{CO}.\text{COOH} \\
| \\
\text{CH}.\text{COOH} \\
| \\
\text{CH}_2.\text{COOH} \\
\text{oxalosuccinic} \\
\text{acid}
\end{array}
\rightleftharpoons
\begin{array}{c}
\text{CO}.\text{COOH} \\
| \\
\text{CH}_2 \\
| \\
\text{CH}_2.\text{COOH} \\
\alpha\text{-oxoglutaric} \\
\text{acid}
\end{array}
+ \text{CO}_2
$$

$$(4)$$

The compound thus formed is α-oxoglutaric acid, which is an α-ketoacid analogous to pyruvic acid. In just the same way as pyruvic acid was oxidized to acetyl coenzyme A, so α-oxoglutaric acid is oxidized to succinyl coenzyme A. The reaction involves thiamine pyrophosphate, lipoic acid and coenzyme A, and reduces NAD; the enzyme is α-oxoglutarate dehydrogenase.

$$\begin{array}{l} \text{CO.COOH} \\ | \\ \text{CH}_2 + \text{CoASH} + \text{NAD} \rightleftharpoons \text{CH}_2.\text{CO.S.CoA} + \text{CO}_2 + \text{NADH}_2 \quad (5) \\ | \qquad\qquad\qquad\qquad\qquad | \\ \text{CH}_2.\text{COOH} \qquad\qquad\qquad \text{CH}_2.\text{COOH} \\ \text{α-oxoglutaric acid} \qquad\quad \text{succinyl coenzyme A} \end{array}$$

Succinyl coenzyme A is analogous to the acetyl coenzyme A formed in reaction (1). It contains a high-energy acyl mercaptide bond (see Table 7.1), and this can now be used for the synthesis of ATP – though in fact the compound first formed appears to be GTP rather than ATP. The reaction is catalysed by a thiokinase and produces succinic acid and free coenzyme A.

$$\begin{array}{l} \text{CH}_2.\text{CO.S.CoA} \qquad\qquad \text{CH}_2.\text{COOH} \\ | \qquad\qquad +\text{GDP}+\text{P}_i+\text{H}_2\text{O} \rightleftharpoons | \qquad\qquad +\text{GTP}+\text{CoASH} \quad (6) \\ \text{CH}_2.\text{COOH} \qquad\qquad\qquad\qquad \text{CH}_2.\text{COOH} \\ \text{succinyl} \qquad\qquad\qquad\qquad\qquad\quad \text{succinic} \\ \text{coenzyme A} \qquad\qquad\qquad\qquad\qquad \text{acid} \end{array}$$

$$\text{GTP}+\text{ADP} \rightleftharpoons \text{GDP}+\text{ATP} \qquad\qquad (6a)$$

Since acetyl coenzyme A also contained an energy-rich acyl mercaptide bond, it is at first sight surprising that it too was not used for the synthesis of GTP or ATP. The reason is that the energy was needed for the condensation with oxaloacetic acid to yield citric acid (reaction (2)).

The succinic acid produced in reaction (6) is oxidized by means of succinate dehydrogenase. This enzyme contains FAD as prosthetic group (see p. 76) and does not use NAD as a coenzyme.

$$\begin{array}{l} \text{CH}_2.\text{COOH} \qquad\quad \text{CH.COOH} \\ | \qquad\qquad \rightleftharpoons \quad \| \qquad\qquad +\text{2H} \qquad\qquad (7) \\ \text{CH}_2.\text{COOH} \qquad\quad \text{CH.COOH} \\ \text{succinic} \qquad\qquad\quad \text{fumaric} \\ \text{acid} \qquad\qquad\qquad\; \text{acid} \end{array}$$

Fumaric acid accepts water in a reaction catalysed by a hydratase. The malic acid that results is oxidized by NAD with malate dehydrogenase as the enzyme.

$$CH.COOH \atop \| \hspace{1.2em} +H_2O \rightleftharpoons \atop CH.COOH \hspace{3em}$$

$$\begin{array}{l} CH.COOH \\ \| \hspace{2em} +H_2O \rightleftharpoons \\ CH.COOH \end{array} \quad \begin{array}{l} CHOH.COOH \\ | \\ CH_2.COOH \end{array} \qquad (8)$$

fumaric acid   malic acid

$$\begin{array}{l} CHOH.COOH \\ | \hspace{2em} +NAD \rightleftharpoons \\ CH_2.COOH \end{array} \quad \begin{array}{l} CO.COOH \\ | \hspace{2em} +NADH_2 \\ CH_2.COOH \end{array} \qquad (9)$$

malic acid   oxaloacetic acid

The net effect of the last three reactions is to oxidize $-CH_2-$ to $-CO-$. This is an important conversion, and we shall see (p. 104) that it occurs in the oxidation of fatty acids by an analogous sequence of three reactions.

The final reaction produces oxaloacetic acid, which can react with acetyl coenzyme A again (2). It is for this reason that the reactions just described are collectively called the Krebs *cycle*.

During the course of the reactions mentioned in this chapter, pyruvic acid is completely oxidized. The three carbon atoms of pyruvic acid are lost as carbon dioxide in reactions (1), (4) and (5). Five pairs of hydrogen atoms are released, in reactions (1), (4), (5), (7) and (9). Thus the overall reaction is formally

$$CH_3.CO.COOH + 3H_2O \longrightarrow 3CO_2 + 5(H_2).$$

Of the five pairs of hydrogen atoms, four pairs are passed to NAD. Each pair can give rise to three molecules of ATP (see p. 80). The remaining pair, produced in reaction (7), is passed to FAD and can give rise to two molecules of ATP (see p. 80). There is also a gain of one ATP (actually GTP in the first instance) in reaction (6). Thus the oxidation of each molecule of pyruvic acid can yield fifteen molecules of ATP.

We can now calculate the yield of ATP from the breakdown of glucose in aerobic conditions. We saw in the previous chapter that the glycolytic pathway produces two molecules of pyruvic acid and yields two molecules of ATP and two of $NADH_2$. In aerobic conditions these two $NADH_2$ can themselves yield six ATP because the respiratory chain will be operating. If each molecule of pyruvic acid gives fifteen ATP the total yield from the oxidation of glucose will be thirty-eight ATP. This calculation is slightly academic and must not be taken completely seriously, since the theoretical yield of ATP cannot be realized in all physiological conditions; nonetheless it gives an indication of what yield can be expected, and it provides a striking contrast with the yield of two ATP under anaerobic conditions.

Replenishment of Krebs cycle intermediates

The Krebs cycle is responsible for the oxidation of the acetyl group of acetyl coenzyme A, and is therefore the final common path of oxidation of carbohydrates and also (see next chapter) of fats. But the Krebs cycle intermediates have other functions too. We shall see later (p. 136) that α-oxoglutaric acid can be converted to the amino acid glutamic acid, which itself is not only a constituent of proteins but also able to give rise to other amino acids. Oxaloacetic acid can be converted to the amino acid aspartic acid (p. 139), and this can give rise to further amino acids and also to pyrimidines. Succinyl coenzyme A is involved in the synthesis of pyrroles, which are constituents of haem and of chlorophyll.

Were it not for these important synthetic side-reactions, only catalytic quantities of Krebs cycle intermediates would be required to function in the oxidation of acetyl groups. But since these intermediates are constantly siphoned off for biosynthetic reactions, there must be a mechanism for replenishing the constituents of the Krebs cycle. Without such a mechanism, the terminal oxidation of carbohydrates and fats would soon fail. In part, the Krebs cycle can be replenished by the reverse of the reactions that have just been mentioned – e.g. α-oxoglutaric acid not only gives rise to glutamic acid but also can be formed from glutamic acid. But in conditions where biosynthetic reactions predominate, there must be another means of replenishing the Krebs cycle intermediates. This is provided by an ATP-dependent carboxylation of pyruvic acid catalysed by pyruvate carboxylase:

$$CH_3.CO.COOH + CO_2 + ATP \rightleftharpoons \begin{array}{c} CO.COOH \\ | \\ CH_2.COOH \end{array} + ADP + P_i$$

pyruvic acid   oxaloacetic acid

The existence of this reaction has an extremely important implication. It requires that the glycolytic pathway be functioning in order to provide pyruvic acid, and thus in the absence of active carbohydrate metabolism replenishment of the Krebs cycle intermediates will be difficult or impossible. We shall return to this point in discussing the oxidation of fatty acids on p. 106.

12 Synthesis of ATP – fat breakdown

In many organisms, a large fraction of the ATP produced comes from the oxidation of fats. Fats are esters of glycerol, having the general formula

$$CH_2O.COR$$
$$|$$
$$CHO.COR'$$
$$|$$
$$CH_2O.COR''$$

and their hydrolysis is readily catalysed by the action of lipases to produce glycerol and fatty acids.

$$
\begin{array}{lll}
CH_2O.COR & CH_2OH & RCOOH \\
| & | & + \\
CHO.COR' \ +3H_2O \rightleftharpoons & CHOH \ + & R'COOH \\
| & | & + \\
CH_2O.COR'' & CH_2OH & R''COOH
\end{array}
$$

We have already seen (p. 94) how glycerol can be metabolized by the reactions of the Embden–Meyerhof pathway. We shall now consider the oxidation of fatty acids.

In principle fats can contain many different fatty acids, but in practice the naturally occurring types of fatty acid are fairly few. They nearly always have an even number of carbon atoms; two of the commonest are palmitic acid ($C_{15}H_{31}COOH$) and stearic acid ($C_{17}H_{35}COOH$). We shall illustrate the oxidation of fatty acids by taking stearic acid as an example. For the present purpose it is convenient to write it as

$$CH_3.(CH_2)_{12}.CH_2.CH_2.CH_2.CH_2.COOH.$$

Before it can be oxidized, the fatty acid has to be activated by reaction with coenzyme A. ATP is used up, being converted to AMP rather than to ADP and releasing inorganic pyrophosphate:

$$CH_3.(CH_2)_{12}.CH_2.CH_2.CH_2.CH_2.COOH + CoASH + ATP \rightleftharpoons$$
$$CH_3.(CH_2)_{12}.CH_2.CH_2.CH_2.CH_2.CO.S.CoA + AMP + (P\!-\!P)_i.$$

This reaction is catalysed by a thiokinase and has an equilibrium constant reasonably near to 1; but the pyrophosphate produced is rapidly degraded owing to the presence of another enzyme, pyrophosphatase:

$$(P-P)_i + H_2O \rightleftharpoons 2P_i.$$

Since the concentration of water in the cell is much higher than that of inorganic phosphate, there is in practice no chance of a net reversal of this latter reaction and thus of a net resynthesis of ATP from AMP and pyrophosphate. We may therefore regard the formation of fatty acyl coenzyme A from fatty acid as effectively irreversible. We may recall that glucose too (and other sugars and glycerol) have first to be activated, by reactions that are effectively irreversible, before being further metabolized.

Stearoyl coenzyme A is now broken down by a series of reactions that are together called β-oxidation. In the course of this series of reactions, the β-carbon atom of the acyl coenzyme A is gradually oxidized to yield a β-keto acyl coenzyme A.

In the first reaction of β-oxidation, a pair of hydrogen atoms is removed from stearoyl coenzyme A to give the corresponding α:β-unsaturated compound. The enzyme for this oxidation is fatty acyl coenzyme A dehydrogenase; it contains FAD as a prosthetic group instead of using NAD:

$$CH_3.(CH_2)_{12}.CH_2.CH_2.CH_2.CH_2.CO.S.CoA \rightleftharpoons$$
$$CH_3.(CH_2)_{12}.CH_2.CH_2.CH{=}CH.CO.S.CoA.$$

Next, the unsaturated or enoyl coenzyme A accepts water in a reaction catalysed by enoyl hydratase. This enzyme adds water specifically to form the β- (not the α-) hydroxy acyl coenzyme A.

$$CH_3.(CH_2)_{12}.CH_2.CH_2.CH{=}CH.CO.S.CoA + H_2O \rightleftharpoons$$
$$CH_3.(CH_2)_{12}.CH_2.CH_2.CHOH.CH_2.CO.S.CoA.$$

This product is now oxidized again, a β-hydroxy fatty acyl coenzyme A dehydrogenase passing a pair of hydrogen atoms to NAD.

$$CH_3.(CH_2)_{12}.CH_2.CH_2.CHOH.CH_2.CO.S.CoA + NAD \rightleftharpoons$$
$$CH_3.(CH_2)_{12}.CH_2.CH_2.CO.CH_2.CO.S.CoA + NADH_2.$$

This series of three reactions is strikingly similar to the three reactions in the Krebs cycle by which succinic acid is oxidized to oxaloacetic acid (p. 99). The first is a dehydrogenation to produce a double bond, catalysed by an enzyme that contains FAD. The second is a hydration, and the third is an oxidation of —CHOH— to —CO— catalysed by an enzyme that uses NAD.

Now that the β-carbon atom has been completely oxidized, the resulting β-keto acyl coenzyme A can be split with another molecule of coenzyme A. This reaction is catalysed by a β-keto thiolase.

$$CH_3.(CH_2)_{12}.CH_2.CH_2.CO.CH_2.CO.S.CoA + CoASH \rightleftharpoons$$
$$CH_3.(CH_2)_{12}.CH_2.CH_2.CO.S.CoA + CH_3.CO.S.CoA.$$

The products are acetyl coenzyme A and the fatty acyl coenzyme A two carbon atoms shorter than the original compound, namely palmitoyl coenzyme A. Acetyl coenzyme A enters the Krebs cycle in the normal way (p. 98) and palmitoyl coenzyme A re-enters the β-oxidation pathway.

$$CH_3.(CH_2)_{12}.CH_2.CH_2.CO.S.CoA$$
$$\downarrow \text{dehydrogenase}$$
$$CH_3.(CH_2)_{12}.CH{=}CH.CO.S.CoA$$
$$\downarrow \text{hydratase}$$
$$CH_3.(CH_2)_{12}.CHOH.CH_2.CO.S.CoA$$
$$\downarrow \text{dehydrogenase}$$
$$CH_3.(CH_2)_{12}.CO.CH_2.CO.S.CoA$$
$$\downarrow \text{thiolase}$$
$$CH_3.(CH_2)_{12}.CO.S.CoA + CH_3.CO.S.CoA$$

In this way fatty acids (in the form of their acyl coenzyme A derivatives) are broken down to several acetyl coenzyme A units, each of which is then oxidized through the Krebs cycle. Stearoyl coenzyme A, for example, undergoes the β-oxidation sequence eight times and yields nine molecules of acetyl coenzyme A.

We can use these figures to calculate the yield of ATP from the complete oxidation of stearic acid. Each sequence of β-oxidation involves the passage of one pair of hydrogen atoms to FAD (giving rise by oxidation to two molecules of ATP) and one pair to NAD (giving rise by oxidation to three molecules of ATP). Thus the β-oxidation sequences for stearoyl coenzyme A should yield

$8 \times 5 = 40$ ATP. The oxidation of nine acetyl groups (entering the Krebs cycle as acetyl coenzyme A) should yield $9 \times 12 = 108$ ATP (see p. 100). Thus the complete oxidation of the stearoyl group yields about 148 ATP molecules.

However, stearoyl coenzyme A was formed from stearic acid in a reaction in which ATP was broken down to AMP, so that the net yield would be diminished by two high energy bonds. Notice that the first β-oxidation sequence produced palmitoyl coenzyme A, not palmitic acid; so that each fatty acyl coenzyme A derived by cleavage with β-ketothiolase can in turn undergo β-oxidation: only *one* activation with ATP is needed for the entire fatty acid.

We may therefore say that the complete oxidation of stearic acid yields about 146 ATP. As before, this calculation must be taken with a grain of salt, but it is instructive to compare the yield from the oxidation of stearic acid (which contains 18 carbon atoms) with that from the oxidation of three molecules of glucose (also containing 18 carbon atoms). Three molecules of glucose would yield $3 \times 38 = 114$ molecules of ATP. The considerably greater yield from the oxidation of stearic acid reflects the fact that a fatty acid is much more reduced than a sugar, i.e. it contains proportionately more hydrogen atoms to be oxidized.

Formation of ketone bodies

In starvation, or in certain pathological conditions (such as diabetes) in which the breakdown of carbohydrates is impaired, the animal tends to turn to fat oxidation to supply its needs of ATP. As a result there is a large flow of acetyl coenzyme A into the Krebs cycle, requiring oxaloacetic acid for its further metabolism. Now we have mentioned (p. 101) that the cycle is an important source of biosynthetic intermediates, and that the supply of oxaloacetic acid is therefore liable to run short unless replenished. But the carboxylation reaction (p. 101) that replenishes the level of oxaloacetic acid can only proceed if there is a supply of pyruvic acid arising from the metabolism of carbohydrates. Yet it is just when there is maximum demand for oxaloacetic acid that the metabolism of carbohydrate has, in fact, failed.

The result is that the Krebs cycle is unable to metabolize the acetyl group of acetyl coenzyme A at a sufficient rate to prevent the concentration of acetyl coenzyme A from rising. Under these conditions the penultimate compound in fatty-acid metabolism, *acetoacetyl* coenzyme A, is deacylated to give acetoacetic acid instead of reacting with coenzyme A to give two molecules of acetyl coenzyme A. Reduction of acetoacetic acid gives β-hydroxybutyric acid:

$$CH_3COCH_2COOH + NADH_2 \rightleftharpoons CH_3CHOHCH_2COOH + NAD$$

acetoacetic acid β-hydroxybutyric acid

and decarboxylation of acetoacetic acid gives acetone:

$$CH_3COCH_2COOH \rightleftharpoons CO_2 + CH_3COCH_3.$$

acetoacetic acid   acetone

These three compounds – acetoacetic acid, β-hydroxybutyric acid and acetone – are known as ketone bodies, and their presence is characteristic of the *ketosis* observed in starvation, diabetes and some other conditions.

13 Synthesis of ATP and NADPH$_2$ – the pentose phosphate pathway

In Chs. 10 and 11 we have seen how glucose and other sugars are metabolized via the Embden–Meyerhof pathway and the Krebs cycle. These pathways account for the bulk of carbohydrate metabolism in most organisms. There is, however, an alternative pathway for the oxidation of sugars that we must now consider. It is sometimes called the Warburg–Dickens pathway, sometimes the hexose monophosphate 'shunt' and sometimes (for reasons that will become obvious) the pentose phosphate pathway.

Like glycolysis, this pathway employs glucose-6-phosphate (p. 89) as its starting material, but unlike glycolysis the pentose phosphate pathway is not a fermentation. We recall that the essential feature of a fermentation is that the molecule undergoes neither net oxidation nor net reduction; here, on the contrary, the very first reaction is an oxidation, and the pair of hydrogen atoms removed from glucose-6-phosphate is not at any stage returned to the molecule. The enzyme glucose-6-phosphate dehydrogenase passes these hydrogen atoms to NADP; we shall consider their fate later.

glucose-6-phosphate   6-phosphogluconolactone

The resulting lactone is unstable and can hydrolyse at a measurable rate in the absence of an enzyme. However, there is, in fact, a lactonase that increases the rate of reaction.

[108]

6-phosphogluconolactone   6-phosphogluconic acid

Next another oxidation occurs. The enzyme, 6-phosphogluconate dehydrogenase, passes two hydrogen atoms to NADP, and the very unstable acid that results decarboxylates spontaneously. This reaction is similar to the oxidation of isocitric acid to another unstable intermediate (oxalosuccinic acid) that also decarboxylates at once (p. 98).

6-phosphogluconic acid   ribulose-5-phosphate

The ribulose-5-phosphate that is formed in this reaction can be converted by an epimerase to xylulose-5-phosphate, or by an isomerase to ribose-5-phosphate.

ribulose-5-phosphate   xylulose-5-phosphate

$$
\begin{array}{ccc}
\text{CH}_2\text{OH} & \text{CHO} \\
| & | \\
\text{CO} & \text{HOCH} \\
| & | \\
\text{HCOH} \;\rightleftharpoons\; & \text{HCOH} \quad \text{i.e.} \\
| & | \\
\text{HCOH} & \text{HCOH} \\
| & | \\
\text{CH}_2\text{O}\,\textcircled{P} & \text{CH}_2\text{O}\,\textcircled{P}
\end{array}
$$

ribulose-5-phosphate   ribose-5-phosphate

Before discussing the further metabolism of these pentose phosphates, we may stop to consider what are the particular functions of the reactions we have described so far. They appear to represent an alternative means for the breakdown of glucose, different from the Embden–Meyerhof–Krebs pathways. But why should a second route be needed for glucose breakdown? We have already seen that the Embden–Meyerhof pathway permits the degradation of glucose in anaerobic conditions, and that the same pathway combined with the Krebs cycle results in the total oxidation of glucose to carbon dioxide and water. As far as the formation of ATP is concerned, these routes provide adequate means of breaking down glucose; so it seems surprising that another pathway should exist alongside the Embden–Meyerhof–Krebs route.

To find out the significance of the reactions leading from glucose-6-phosphate to the pentose phosphates, let us look more closely at the compounds they produce.

In the first place, these reactions lead to pentoses and in particular to ribose. Ribose is a constituent of such coenzymes as NAD and ATP, and, of course, of the nucleosides in RNA. From the nucleoside triphosphates (ATP, CTP, etc.) that are used in the synthesis of RNA, deoxynucleoside triphosphates (deoxyATP, deoxyCTP, etc.) can be formed; these are the building blocks for the synthesis of DNA (deoxyribonucleic acid). Some of these conversions will be described in Ch. 19.

Secondly, the oxidation of glucose-6-phosphate yields not only pentose phosphates but also $NADPH_2$. We saw in Ch. 8 that $NADH_2$ is for the most part reoxidized by the respiratory chain and that this oxidation yields ATP. On the other hand, $NADPH_2$

is used primarily as a reducing agent (see Ch. 7). Just as the complete breakdown of large molecules (e.g. lipids and poly-saccharides) involves their oxidation, so the synthesis of these molecules from simple precursors requires reduction. Fatty acids, for example, are synthesized starting from acetyl coenzyme A (Ch. 16); since the acetyl group is CH_3CO—, whereas fatty acids have long chains of the type $CH_3CH_2CH_2CH_2$ ..., it is evident that reduction as well as condensation is required in the synthesis: we shall see in detail (Ch. 16) how this reduction by $NADPH_2$ takes place. $NADPH_2$ is also used in the synthesis of steroids and in other important reductions.

A tissue that is actively synthesizing fats or steroids will require large quantities of $NADPH_2$. We have seen that the oxidation of glucose-6-phosphate to ribulose-5-phosphate reduces two mole-cules of NADP to $NADPH_2$, but two molecules of $NADPH_2$ suffice only for the reduction of one acetyl group to —CH_2CH_2— in a fatty acid (see p. 132). To synthesize (for example) stearic acid requires sixteen molecules of $NADPH_2$, i.e. the oxidation of eight molecules of glucose-6-phosphate to eight molecules of ribulose-5-phosphate.

We have just pointed out that a supply of pentoses is extremely important; but there can be too much of a good thing, and for a tissue rapidly synthesizing fats or steroids the quantity of ribulose-5-phosphate produced as a necessary by-product of reducing NADP may well be an embarrassment. There is, however, a means of converting pentose back to hexose by shuffling around the carbon atoms of various sugars and thus forming five hexoses from six pentoses. These reactions for the interconversion of sugars form the second part of the Warburg–Dickens pathway, and we shall see (Ch. 15) that similar reactions are used in the interconversion of sugars after the fixation of carbon dioxide in photosynthesis.

Two special enzymes are involved in these interconversions. Transketolase catalyses the transfer of the $CH_2OH.CO$— group from a ketose to an aldose, and transaldolase the transfer of the $CH_2OH.CO.CHOH$— group, also from a ketose to an aldose. Thus the transketolase transfers a two-carbon fragment and the transaldolase a three-carbon fragment, and between them they can accomplish the conversion of three molecules of pentose phosphate into two molecules of hexose phosphate and one of triose phosphate:

$$\mathbf{C_5} + \mathbf{C_5} \xrightleftharpoons[]{\text{transketolase}} C_7 + C_3$$

$$C_7 + C_3 \xrightleftharpoons[]{\text{transaldolase}} C_4 + C_6$$

$$\mathbf{C_5} + C_4 \xrightleftharpoons[]{\text{transketolase}} C_3 + C_6$$

(Bold type signifies the original pentose phosphate molecules.)

Writing the reactions in more detail gives:

xylulose-5-Ⓟ + ribose-5-Ⓟ ⇌ sedoheptulose-7-Ⓟ
$$\qquad\qquad\qquad\qquad\qquad\qquad + \text{3-phosphoglyceraldehyde}$$

sedoheptulose-7-Ⓟ + 3-phosphoglyceraldehyde ⇌
$$\qquad\qquad\qquad\qquad\quad \text{erythrose-4-Ⓟ} + \text{fructose-6-Ⓟ}$$

xylulose-5-Ⓟ + erythrose-4-Ⓟ ⇌ 3-phosphoglyceraldehyde
$$\qquad\qquad\qquad\qquad\qquad\qquad + \text{fructose-6-Ⓟ}$$

If the whole of this sequence is carried out twice, the six molecules of pentose phosphate are converted into four molecules of fructose-6-phosphate and two molecules of 3-phosphoglyceraldehyde. It is formally possible for the two molecules of 3-phosphoglyceraldehyde to give rise to fructose-1;6-diphosphate, by the reverse of reactions (4) and (5) of the Embden–Meyerhof pathway (p. 91); fructose-1;6-diphosphate can then be cleaved by its specific phosphatase (p. 90) to give fructose-6-phosphate. Thus, as we mentioned above, the sequence provides for the conversion of six molecules of pentose phosphate into five molecules of hexose phosphate.

We have seen (p. 109) that the original pentose phosphate that derived from the oxidation of glucose-6-phosphate was ribulose-5-phosphate, which then became converted into xylulose-5-phosphate and ribose-5-phosphate. In a similar way, the fructose-6-phosphate which is the product of the rearrangements by transketolase and transaldolase can be converted into glucose-6-phosphate by hexose phosphate isomerase (p. 90). It is thus possible to summarize the reactions as

$$\text{6 ribulose-5-phosphate} \xrightleftharpoons[]{} \text{5 glucose-6-phosphate} + P_i$$

and a flow-sheet illustrating these conversions is given in Fig. 13.1.

We should emphasize that, although all these reactions can occur and, in some circumstances, certainly do, the metabolic

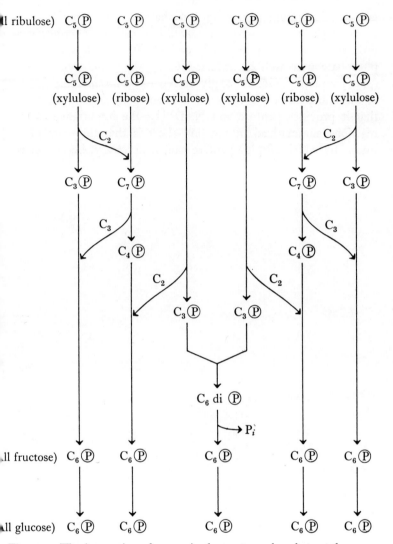

Fig. 13.1. The interaction of sugars in the pentose phosphate pathway.

conversions are not nearly as formal as they appear here. Frequently both the Embden–Meyerhof and the Warburg–Dickens
pathways are operating in the same cell at the same time. Thus the
fructose-6-phosphate formed by the transketolase and transaldo-

lase reactions from ribulose-5-phosphate may well not be converted back into glucose-6-phosphate but instead undergo glycolysis. Similarly the 3-phosphoglyceraldehyde derived from pentose phosphate may well be oxidized to 1;3-diphosphoglyceric acid (p. 91) or perhaps give rise to glycerol (p. 94). The chief importance of the first part of the Warburg–Dickens pathway appears to be that it provides pentose and $NADPH_2$; the importance of the reactions summarized in the flow-sheet is that they provide a means by which surplus pentose can, if necessary, be converted back to hexose.

14 Synthesis of ATP and NADPH₂ – the light reaction of photosynthesis

In the last four chapters we have seen how heterotrophic organisms (those that live by breaking down organic compounds) derive the bulk of their ATP and $NADPH_2$. But heterotrophic organisms depend ultimately on autotrophic organisms, which are capable of synthesizing organic compounds from simple inorganic materials. It is the ability of autotrophic organisms to fix carbon dioxide into sugars on which all life on the earth depends. In this chapter and the next we shall describe in outline how this fixation occurs.

This chapter is devoted to studying how the prerequisites of carbon dioxide fixation are provided. If we consider the conversion

$$CO_2 \longrightarrow (CH_2O)_n$$

where $(CH_2O)_n$ represents a carbohydrate (e.g. glucose), it is clear that there are two prerequisites. The reaction involves the building up of a very simple molecule into a larger molecule, and therefore requires ATP. It also involves reduction, and therefore requires a reducing agent (which turns out to be $NADPH_2$). Now the fixation of carbon dioxide into carbohydrates is a process largely confined to green plants and certain pigmented micro-organisms, and it is dependent on the energy of sunlight. So the problem that we have to consider in this chapter is, how can light energy be used to provide ATP and to reduce NADP?

The processes by which ATP is synthesized and NADP reduced are often known as the 'light reaction' of photosynthesis, and they are absolutely dependent on the presence of characteristic photosynthetic pigments. These pigments are confined in plants to the structures known as chloroplasts (see p. 67), which can carry out the light reaction even after isolation from the plant. The characteristic pigments of the chloroplasts are various chlorophylls (usually identified by their absorption spectra) and carotenoids, and in addition there are special cytochromes that occur only in plants.

The majority of the chlorophyll molecules in plants belong to the molecular species chlorophyll *a*, and these tend to be organized into arrays of several hundred molecules. Many of these arrays are associated with one molecule of a specialized chlorophyll molecule called P 700 and with a molecule of the specialized plant cytochrome called cytochrome *f*. A complex of this sort forms one unit of a system called 'pigment system I', which is believed to be involved in the primary photochemical event by which ATP is produced.

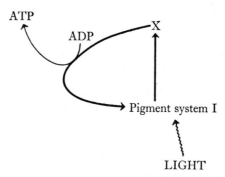

Fig. 14.1. The synthesis of ATP in the chloroplast. Bold arrows represent flow of electrons.

Although the details of the light reaction in photosynthesis are by no means agreed, there is evidence to support some such scheme as that set out in Fig. 14.1. Light falling on any of the chlorophyll *a* molecules in pigment system I causes an electron from the associated chlorophyll P 700 to be promoted to a higher energy level. This electron is donated to an unidentified substance X, which we may now call X^-. The electron can be transferred from X^- to an electron transport chain. This chain consists of cytochromes and other electron carriers, and it is probably quite similar to the electron transport chain of oxidative phosphorylation. At all events, during the passage of an electron along the chain ATP is synthesized from ADP. The final acceptor of the electron, however, is not oxygen but the pigment system I itself, which had been left electron-deficient by the light-induced event. Thus in this system ATP is produced by a cyclical flow of electrons, and the process is sometimes called 'cyclic photophosphorylation'.

This description can account for ATP synthesis, but it does not explain the formation of the other prerequisite of carbon dioxide fixation, $NADPH_2$. For the reduction of NADP (as well as for further synthesis of ATP) a rather more complicated series of reactions appears to take place: this is illustrated in Fig. 14.2. Once

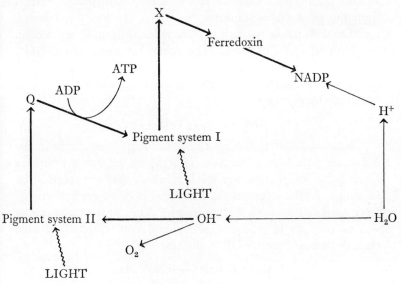

Fig. 14.2. The synthesis of ATP and the reduction of NADP in the chloroplast.

again an electron in pigment system I is promoted to a higher energy level by light and passed to X; but X^-, instead of donating an electron to the electron transport chain, reduces the iron-containing protein ferredoxin which in turn passes an electron to NADP. However, pigment system I is now deficient in an electron, and in order for it to be reduced again a second photochemical event is necessary. This takes place in 'pigment system II', which is a complex containing the specialized chlorophyll called P670. In response to light this chlorophyll loses an electron to an acceptor known as Q, which in turn transfers an electron to pigment system I. Just as before, the electron is passed along an electron transport chain, and this passage is accompanied by the synthesis of ATP.

5

As a result of these events, NADP has acquired an electron and pigment system II has lost one; the intermediate carriers have returned to their previous state. Now in order to form $NADPH_2$ from NADP, not only electrons but also H^+ ions are required; moreover, in order to restore pigment system II to its normal state another electron is required. Both of these requirements are met from the products of ionization of water. H^+ ions from water, together with electrons arising via ferredoxin from X, are used in the formation of $NADPH_2$. The other ion from water, OH^-, donates an electron to the electron-deficient pigment system II, and as a result becomes converted to water and molecular oxygen. This may be written formally as

$$4OH^- \longrightarrow 2H_2O + O_2 + 4e^-.$$

We can summarize the light reaction of photosynthesis in the following way. The absorption of light by pigments promotes electrons to a higher energy level, at which they are received by acceptors. As the electrons fall back again to their original energy level, they can do work in synthesizing ATP, in the same sort of way as they do in oxidative phosphorylation. Alternatively the electrons can reduce NADP in the reaction

$$NADP + 2e + 2H^+ \longrightarrow NADPH_2.$$

The H^+ needed for this reaction is derived from the ionization of water, and the corresponding OH^- loses an electron as we have seen above.

This description is simplified, and (because our understanding of the light reaction is still very incomplete) it may possibly be incorrect. It does, however, account for the characteristic features of photosynthesis in green plants: the synthesis of ATP and the reduction of NADP, and the concomitant evolution of molecular oxygen.

15 Uses of ATP and NADPH₂ – the synthesis of polysaccharides

Armed with ATP and $NADPH_2$, the synthesis of which we discussed in the last chapter, plants are capable of fixing carbon dioxide into sugars and thus, ultimately, of synthesizing starch. In this chapter we shall describe some of the details of these processes, which are often called the 'dark reaction' of photosynthesis.

We can write the outline of carbon dioxide fixation in this way.

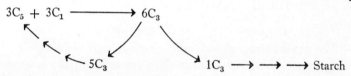

The key reaction is that in which carbon dioxide reacts with a pentose to yield two molecules of triose. If we consider this reaction as multiplied by three, the yield is six molecules of triose. Of these six, one molecule is built up to glucose and eventually to starch; the remaining five are rearranged to form three molecules of the pentose which can again accept carbon dioxide. These processes are sometimes called the Calvin cycle.

The Calvin cycle

The pentose with which carbon dioxide reacts is the diphosphate derivative of ribulose (see pp. 65 and 70). The enzyme is carboxydismutase, and the reaction yields two molecules of 3-phosphoglyceric acid.

$$
\begin{array}{l}
CH_2O\text{℗} \\
|\\
CO \\
|\\
HCOH \\
|\\
HCOH \\
|\\
CH_2O\text{℗}
\end{array}
+CO_2+H_2O \rightleftharpoons
\begin{array}{l}
CH_2O\text{℗} \\
|\\
CHOH \\
|\\
COOH \\
+\\
COOH \\
|\\
CHOH \\
|\\
CH_2O\text{℗}
\end{array}
\qquad (1)
$$

$\qquad$ ribulose-1;5-diphosphate $\qquad$ 2 × 3-phosphoglyceric acid

The 3-phosphoglyceric acid is now phosphorylated by ATP in the presence of a kinase, and the resulting 1;3-diphosphoglyceric acid is reduced by $NADPH_2$ to 3-phosphoglyceraldehyde. The enzyme is triose phosphate dehydrogenase.

$$
\begin{array}{l}
\text{COOH} \\
|\\
\text{CHOH} \\
|\\
\text{CH}_2\text{O}\textcircled{P}
\end{array}
+ \text{ATP} \rightleftharpoons
\begin{array}{l}
\text{COO}\textcircled{P} \\
|\\
\text{CHOH} \\
|\\
\text{CH}_2\text{O}\textcircled{P}
\end{array}
+ \text{ADP} \qquad (2)
$$

3-phosphoglyceric acid   1;3-diphosphoglyceric acid

$$
\begin{array}{l}
\text{COO}\textcircled{P} \\
|\\
\text{CHOH} \\
|\\
\text{CH}_2\text{O}\textcircled{P}
\end{array}
+ \text{NADPH}_2 \rightleftharpoons
\begin{array}{l}
\text{CHO} \\
|\\
\text{CHOH} \\
|\\
\text{CH}_2\text{O}\textcircled{P}
\end{array}
+ \text{NADP} + \text{P}_i \qquad (3)
$$

1;3-diphosphoglyceric acid   3-phosphoglyceraldehyde

In essence, these last two reactions are precisely the reverse of reactions (6) and (7) of the Embden–Meyerhof pathway (p. 91). That pathway is devoted to the breakdown of carbohydrates; photosynthesis is devoted to the formation of carbohydrates. Reaction (6) of the Embden–Meyerhof pathway is an oxidation; reaction (3) of the Calvin cycle is a corresponding reduction (note, however, that it uses $NADPH_2$ (pp. 73 and 117)). Reaction (7) of the Embden–Meyerhof pathway synthesizes ATP; reaction (2) of the Calvin cycle uses ATP.

In reactions (2) and (3) of the Calvin cycle we can therefore see, in part, how ATP and $NADPH_2$, formed in the light reaction, are used in the dark reaction. We shall discover soon (p. 122) that there is a further requirement for ATP in the rearrangement leading to ribulose-1;5-diphosphate.

Before considering the details of this rearrangement, we ought to have a look at the stoichiometry of reactions (1), (2) and (3). It is convenient (see above) to start with three molecules of ribulose diphosphate; thus, multiplying reaction (1) by three and reactions (2) and (3) by six and adding them up, we get equation (4).

$$
\begin{array}{l}
3 \times \text{ribulose-1;5-diphosphate} \\
+ 3\text{CO}_2 \\
+ 3\text{H}_2\text{O} \\
+ 6\text{ATP} \\
+ 6\text{NADPH}_2
\end{array}
\longrightarrow
\begin{array}{l}
6 \times \text{3-phosphoglyceraldehyde} \\
+ 6\text{P}_i \\
+ 6\text{ADP} \\
+ 6\text{NADP}
\end{array}
\qquad (4)
$$

Of these six molecules of 3-phosphoglyceraldehyde, five represent merely a rearrangement of three molecules of ribulose diphosphate, but one represents a net gain to the system. In other words one molecule of triose phosphate on the right-hand side of this sum has been synthesized from the three molecules of carbon dioxide on the left-hand side. It is the achievement of this synthesis that is characteristic of photosynthesis. By contrast the reactions involved in building up sugars and polysaccharides from triose phosphate are essentially similar in photosynthetic and non-photosynthetic organisms; we shall follow these reactions later (p. 124).

The remaining five molecules of triose phosphate can now be rearranged. Many of the reactions involved are similar to those we described in the Warburg–Dickens pathway (Ch. 13) – the trans-ketolase (p. 111) used in that pathway is required here as well (although the transaldolase is not). In addition triose phosphate isomerase (p. 91) is needed to interconvert 3-phosphoglyceralde-hyde and dihydroxyacetone phosphate.

$$
\begin{array}{ccc}
\text{CHO} & & \text{CH}_2\text{OH} \\
| & & | \\
\text{CHOH} & \rightleftharpoons & \text{CO} \\
| & & | \\
\text{CH}_2\text{O}\textcircled{P} & & \text{CH}_2\text{O}\textcircled{P}
\end{array}
\qquad (5)
$$

3-phosphoglyceraldehyde   dihydroxyacetone phosphate

We also need the enzyme aldolase. We saw previously (p. 91) that this enzyme can catalyse the breakdown of fructose-1;6-diphos-phate to two molecules of triose phosphate. The same enzyme can, in fact, catalyse the reverse reaction, and, more generally, it can carry out the aldol condensation between dihydroxyacetone phosphate and various aldehydes.

In outline, the conversions are as follows:

$$\mathbf{C_3} + \mathbf{C_3} \underset{\text{aldolase}}{\rightleftharpoons} C_6, \qquad (6)$$

$$C_6 + \mathbf{C_3} \underset{\text{transketolase}}{\rightleftharpoons} C_4 + C_5, \qquad (7)$$

$$C_4 + \mathbf{C_3} \underset{\text{aldolase}}{\rightleftharpoons} C_7, \qquad (8)$$

$$C_7 + \mathbf{C_3} \underset{\text{transketolase}}{\rightleftharpoons} C_5 + C_5, \qquad (9)$$

(Bold type signifies the original triose phosphate molecules.)

Writing the reactions in more detail gives:

3-phosphoglyceraldehyde + dihydroxyacetone phosphate $\rightleftharpoons$
$$\text{fructose-1;6-diphosphate} \rightleftharpoons \text{fructose-6-phosphate} + P_i, \quad (6)$$

fructose-6-phosphate + 3-phosphoglyceraldehyde $\rightleftharpoons$
$$\text{erythrose-4-phosphate} + \text{xylulose-5-phosphate}, \quad (7)$$

erythrose-4-phosphate + dihydroxyacetone phosphate $\rightleftharpoons$
$$\text{sedoheptulose-1;7-diphosphate} \rightleftharpoons \text{sedoheptulose-7-phosphate} + P_i, \quad (8)$$

sedoheptulose-7-phosphate + 3-phosphoglyceraldehyde $\rightleftharpoons$
$$\text{ribose-5-phosphate} + \text{xylulose-5-phosphate}. \quad (9)$$

Each of the two xylulose-5-phosphate molecules formed (reactions (7) and (9)) can be converted to ribulose-5-phosphate. Similarly, the ribose-5-phosphate produced in reaction (9) can also be converted to ribulose-5-phosphate. The enzymes responsible for these conversions have been mentioned on p. 109. A complete flow-sheet for the interconversions is given in Fig. 15.1.

We may note that two of the above reactions, (6) and (8), involve the hydrolysis of a sugar diphosphate by a specific phosphatase to yield a sugar monophosphate. In consequence, the conversions shown in reactions (6) to (9) give rise to only three molecules of organic phosphate from five molecules of organic phosphate; the remaining two phosphate groups are lost by hydrolysis. In other words, the three molecules of ribulose produced are in the form of ribulose monophosphate. However, the fixation of carbon dioxide (reaction (1)) requires ribulose diphosphate. Each ribulose-5-phosphate must therefore be converted to ribulose-1;5-diphosphate at the expense of ATP, as we discussed on p. 70. The reaction is catalysed by a kinase:

$$\text{ribulose-5-phosphate} + \text{ATP} \rightleftharpoons \text{ribulose-1;5-diphosphate} + \text{ADP}. \quad (10)$$

Now that we have arrived at ribulose-1;5-diphosphate, which can react again with carbon dioxide, we have completed an account of the reactions involved in the synthesis of triose phosphate. We can now look once more at the stoichiometry of the Calvin cycle. Summing reactions (6), (7), (8) and (9) gives:

$$5 \times \text{triose phosphate} \longrightarrow 3 \times \text{pentose phosphate} + 2P_i \quad (11)$$

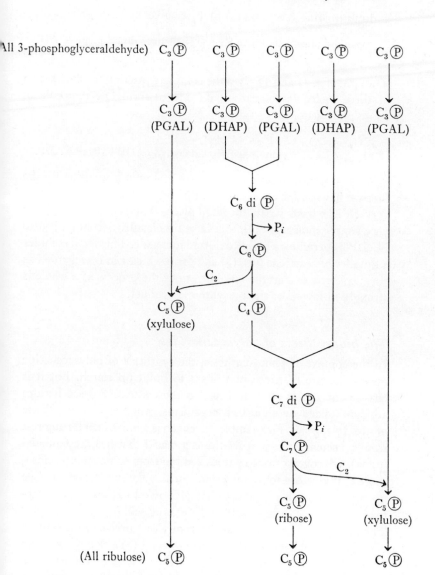

Fig. 15.1. The interconversion of sugars in the Calvin cycle.
PGAL = 3-phosphoglyceraldehyde. DHAP = dihydroxyacetone phosphate.

and multiplying reaction (10) by three gives:

$$3 \times \text{pentose phosphate} + 3\text{ATP} \longrightarrow 3 \times \text{pentose diphosphate} + 3\text{ADP}.$$
$$(12)$$

The sum of 4, 11 and 12 gives the complete reaction for the fixation of three molecules of carbon dioxide to form one molecule of triose phosphate.

$$3CO_2 + 3H_2O + 9\text{ATP} + 6\text{NADPH}_2 \longrightarrow$$
$$\text{3-phosphoglyceraldehyde} + 9\text{ADP} + 8P_i + 6\text{NADP}.$$

This summary reaction represents a simple biosynthesis – the first we have so far considered in this book. Three points emerge very clearly from it. In the first place, it uses ATP to build a very simple molecule up into a larger molecule. Secondly, it uses NADPH_2 to reduce a highly oxidized molecule. Thirdly, it includes a number of reactions ((6), (8) and (10)) which can be regarded as 'irreversible', so that the equilibrium of the process as a whole is strongly in favour of the biosynthetic product.

The biosynthesis of polysaccharides

In the context of photosynthesis, the synthesis of polysaccharide means the use of triose phosphate to build up starch. But it is convenient at this stage to broaden the context. Polysaccharides are synthesized by almost all organisms, whether photosynthetic or not. In animals, for example, glycogen is synthesized during rest from muscular activity from the lactic acid that has accumulated (p. 93). Glycogen can be synthesized too from some amino acids – those that give rise to pyruvic acid, such as alanine or cysteine (see pp. 139 and 143), and those that give rise to Krebs cycle intermediates such as aspartic acid and glutamic acid (pp. 143 and 136). The synthesis of polysaccharide from these precursors proceeds *via* triose phosphate (see reactions (20) and (21) below); and the triose phosphate produced in the Calvin cycle joins the pathway at this stage.

We can for convenience regard the synthesis of polysaccharide as taking place in two stages – the synthesis of glucose-1-phosphate from smaller molecules, and then the polymerization of glucose residues into a polysaccharide chain. The synthesis of glucose-1-phosphate makes use of reactions most of which we have in fact previously mentioned but it is useful to bring them together here. The reactions of polymerization are new ones, which we shall discuss in greater detail.

If we start from lactic acid, the first reaction is an oxidation by NAD with lactate dehydrogenase.

$$
\begin{array}{ccc}
\mathrm{CH_3} & & \mathrm{CH_3} \\
| & & | \\
\mathrm{CHOH} + \mathrm{NAD} \rightleftharpoons & & \mathrm{CO} + \mathrm{NADH_2} \qquad (13) \\
| & & | \\
\mathrm{COOH} & & \mathrm{COOH} \\
\text{lactic acid} & & \text{pyruvic acid}
\end{array}
$$

This is simply the reverse of the reaction given on p. 93; and, just as in the glycolysis pathway the $NADH_2$ needed to reduce pyruvic acid is derived from an oxidation earlier in the pathway, so here $NADH_2$ produced will be used later in the pathway at reaction (19).

In glycolysis, pyruvic acid is formed from phospho-*enol*pyruvic acid (p. 93) but because of the high negative $\Delta G^{0\prime}$ of that reaction (see p. 66), a bypass is needed in order to form phospho-*enol*pyruvic acid from pyruvic acid. This is provided by carboxylation of pyruvic acid to oxaloacetic acid followed by decarboxylation to phospho-*enol*pyruvic acid. The carboxylation reaction was described on p. 101; the decarboxylation is catalysed by an enzyme called phospho-*enol*pyruvate carboxykinase and involves the splitting of GTP.

$$
\mathrm{CH_3.CO.COOH} + \mathrm{CO_2} + \mathrm{ATP} \rightleftharpoons \underset{\substack{| \\ \mathrm{CH_2.COOH}}}{\mathrm{CO.COOH}} + \mathrm{ADP} + \mathrm{P}_i \quad (14)
$$

pyruvic acid   oxaloacetic acid

$$
\underset{\substack{| \\ \mathrm{CH_2.COOH}}}{\mathrm{CO.COOH}} + \mathrm{GTP} \rightleftharpoons \mathrm{CH_2}{=}\mathrm{CO\textcircled{P}.COOH} + \mathrm{GDP} + \mathrm{CO_2} \ (15)
$$

oxaloacetic acid   phospho-*enol*pyruvic acid

Reaction (15) also provides a means by which the carbon atoms of those amino acids that give rise to Krebs cycle intermediates (see Ch. 17) can enter the synthetic sequence.

Phospho-*enol*pyruvic acid accepts water in a reaction catalysed by enolase, and the resulting 2-phosphoglyceric acid is converted by a mutase to 3-phosphoglyceric acid.

$$CH_2{=}CO\textcircled{P}.COOH + H_2O \rightleftharpoons CH_2OH.CHO\textcircled{P}.COOH \quad (16)$$

    phospho-*enol*pyruvic acid   2-phosphoglyceric acid

$$CH_2OH.CHO\textcircled{P}.COOH \rightleftharpoons CH_2O\textcircled{P}.CHOH.COOH \quad (17)$$

     2-phosphoglyceric acid   3-phosphoglyceric acid

3-Phosphoglyceric acid is now phosphorylated by ATP in the presence of a kinase, and the 1;3-diphosphoglyceric acid is reduced by $NADH_2$ in the presence of triose phosphate dehydrogenase.

$$CH_2O\textcircled{P}.CHOH.COOH + ATP \rightleftharpoons CH_2O\textcircled{P}.CHOH.COO\textcircled{P} + ADP$$
$$(18)$$

    3-phosphoglyceric acid   1;3-diphosphoglyceric acid

$$CH_2O\textcircled{P}.CHOH.COO\textcircled{P} + NADH_2 \rightleftharpoons$$
$$CH_2O\textcircled{P}.CHOH.CHO + P_i + NAD \quad (19)$$

    1;3-diphosphoglyceric acid   3-phosphoglyceraldehyde

These last four reactions are simply the reversal of the four corresponding reactions in glycolysis (pp. 91–2).

If the starting point for the synthesis of polysaccharide is lactic acid, the $NADH_2$ needed in reaction (19) is generated by the oxidation reaction (13). If, on the other hand, the starting point for the synthesis is pyruvic acid or oxaloacetic acid, or one of their amino-acid precursors, then the $NADH_2$ must be provided from elsewhere. The synthesis of polysaccharide from one of these latter compounds is another example of the need for reduction in many biosynthetic reactions. Reactions (18) and (19) are parallel with reactions (2) and (3) of the Calvin cycle (p. 120), the only difference being that the reduction in photosynthesis (needed to bring CO_2 to the reduction level of carbohydrate) uses $NADPH_2$ instead of $NADH_2$.

3-Phosphoglyceraldehyde produced in reaction (19) is in equilibrium with dihydroxyacetone phosphate, the equilibrium being catalysed by triose phosphate isomerase. The two triose phosphates can now react to form fructose-1;6-diphosphate in the presence of aldolase.

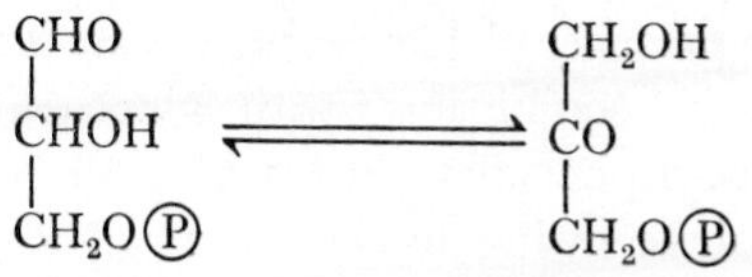

$$\begin{array}{ccc} \text{CHO} & & \text{CH}_2\text{OH} \\ | & & | \\ \text{CHOH} & \rightleftharpoons & \text{CO} \\ | & & | \\ \text{CH}_2\text{O}\,\textcircled{P} & & \text{CH}_2\text{O}\,\textcircled{P} \end{array} \qquad (20)$$

3-phosphoglyceraldehyde   dihydroxyacetone
phosphate

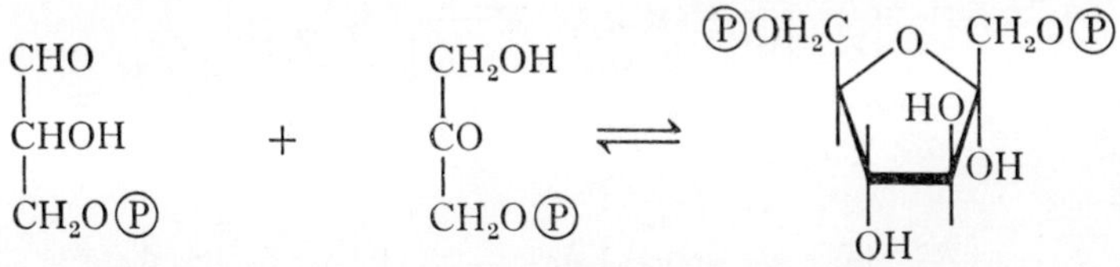

$$\begin{array}{ccc} \text{CHO} & & \text{CH}_2\text{OH} \\ | & & | \\ \text{CHOH} \quad + & & \text{CO} \quad \rightleftharpoons \\ | & & | \\ \text{CH}_2\text{O}\,\textcircled{P} & & \text{CH}_2\text{O}\,\textcircled{P} \end{array}$$

3-phosphoglyceraldehyde   dihydroxyacetone   fructose-1;6-
phosphate   diphosphate

(21)

These two reactions are again reversals of the corresponding
reactions ((5) and (4)) of glycolysis (p. 91). They also correspond to
reactions (5) and (6) of the Calvin cycle (p. 121). Thus the triose
phosphate that was produced by photosynthesis can be regarded
as entering the sequence at this point.

Fructose-1;6-diphosphate is now cleaved in another reaction
that also occurs in the Calvin cycle (p. 122). This reaction, however,
is not the reverse of any reaction in glycolysis; the formation of
fructose-1;6-diphosphate that occurred there (p. 90) was at the
expense of ATP and catalysed by a kinase, whereas the cleavage of
fructose-1;6-diphosphate is an hydrolysis and catalysed by a
phosphatase.

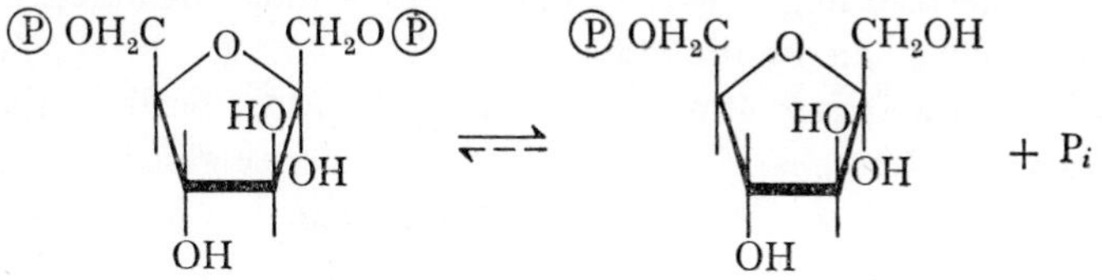

$$\rightleftharpoons \qquad + \text{P}_i$$

(22)

fructose-1;6-diphosphate   fructose-6-phosphate

Owing to the presence of hexose phosphate isomerase fructose-6-phosphate is in equilibrium with glucose-6-phosphate (see p. 64). Glucose-6-phosphate in turn is in equilibrium with glucose-1-phosphate owing to the action of a mutase. The first

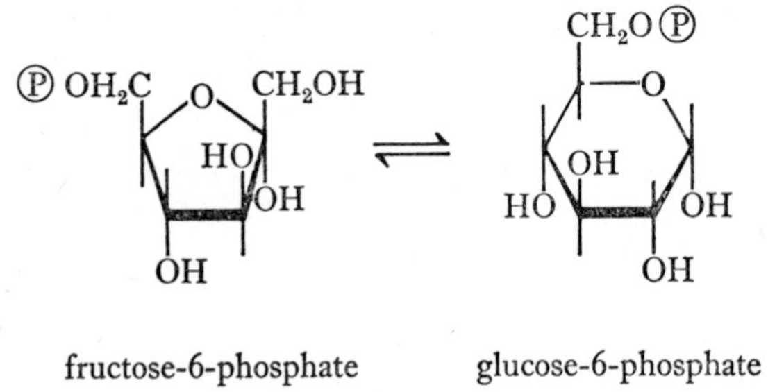

(23)

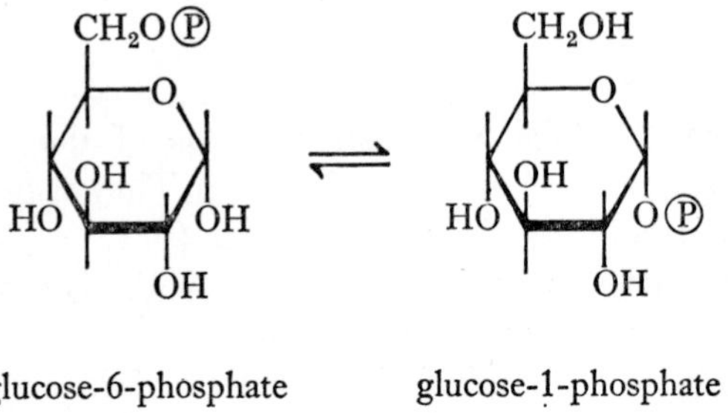

fructose-6-phosphate   glucose-6-phosphate

(24)

glucose-6-phosphate   glucose-1-phosphate

of these reactions is the reverse of reaction (2) of glycolysis (p. 90), and the second has been mentioned in the discussion of glycogen breakdown (p. 95). In the synthesis of glycogen from *glucose*, for example in the liver after absorption of glucose from the gut, glucose is first phosphorylated by the hexokinase reaction (p. 89). The resulting glucose-6-phosphate is then converted to glucose-1-phosphate by the mutase.

We have now arrived at the reactions that lead specifically to the synthesis of polysaccharides. In order to polymerize glucose, it is first necessary to convert the sugar into a form that has a high potential for transfer and can easily be donated to an acceptor molecule in the synthesis of the polymer.

Activation of the glucose molecule is brought about by attaching to it uridine diphosphate (UDP). UDP is analogous to ADP, in that it consists of a base, ribose and two phosphate groups; the base in UDP is uracil, in place of the adenine of ADP. Just as ADP can be further phosphorylated to ATP, so UDP can be further phosphorylated to UTP. However, this phosphorylation occurs not directly (through oxidative phosphorylation, etc.) but by reaction with ATP itself:

$$\text{ATP} + \text{UDP} \rightleftharpoons \text{ADP} + \text{UTP}. \qquad (25)$$

UTP can now react with glucose-1-phosphate, the product of reaction (21), to form UDP-glucose.

| glucose-1-phosphate | UDP-glucose | pyrophosphate |

$$(26)$$

This reaction, catalysed by a pyrophosphorylase, has an equilibrium constant sufficiently near to 1 to allow net formation of product in either direction. However the rapid hydrolysis of pyrophosphate (p. 104) makes the reaction from right to left in practice impossible.

UDP-glucose contains a 'high energy bond', the free energy of hydrolysis of which is about the same as that of ATP, and the glucose can therefore be readily donated to a suitable acceptor. In the synthesis of glucose polymers the acceptor is a primer consisting of a few glucose residues, and the glucose from UDP-glucose becomes transferred to this primer and attached by its C-1, thus lengthening the chain by one unit. The enzyme is a transferase, and UDP is liberated:

$$(\text{glucose})_n + \text{UDP-glucose} \longrightarrow (\text{glucose})_{n+1} + \text{UDP}. \qquad (27)$$

In this way polymers such as amylose (consisting of glucose residues joined by $1 \to 4$, α linkages) (see p. 56) or cellulose (consisting of glucose residues joined by $1 \to 4$, β linkages) (see p. 55) can be built up. Glycogen, which has $1 \to 6$, α linkages as well as $1 \to 4$, α linkages, is made from amylose by a branching enzyme. This enzyme transfers small fragments from the end of the amylose chain to the 6-position of glucose residues.

We can now calculate the requirement for ATP in the synthesis of polysaccharide from lactic acid. Reactions (14), (15) and (18) use up one molecule of ATP each; so three molecules of ATP are used for the synthesis of one molecule of triose phosphate from lactic acid, and therefore six molecules of ATP are used to form hexose monophosphate from two molecules of lactic acid. A further molecule of ATP is used up in reaction (25); thus in total seven ATP molecules are required for each glucose residue originating in lactic acid and polymerized into glycogen. If we start from oxaloacetic acid the requirement will be five ATP molecules since oxaloacetic acid can enter at reaction (15).

What we have said implies that any compound that can provide carbon atoms to the Krebs cycle can be used to synthesize polysaccharide. There is one important sense in which this implication is misleading. When the acetyl group of acetyl coenzyme A enters the Krebs cycle it reacts with oxaloacetic acid to yield citric acid (p. 98). But after one complete revolution of the cycle only oxaloacetic acid remains; the acetyl group has been completely oxidized. Thus acetyl coenzyme A does not add to the stock of intermediates in the Krebs cycle. There is no way in which the cycle can make use of the carbon atoms of the acetyl group for any reaction other than oxidation.

The chief source of acetyl coenzyme A is fatty acids, and it is for these reasons that the fatty acids that we have been discussing cannot be used to provide carbon to replenish the intermediates in the Krebs cycle (p. 101). For exactly the same reasons fatty acids cannot be converted into carbohydrate. Although for other substances the Krebs cycle functions both as the terminal path of oxidation and as a crossroads of metabolic conversions, for fatty acids the cycle can act only as a crematorium.

16 Uses of ATP and NADPH₂ – the synthesis of fats

In Ch. 7 we put forward the idea that the reduced coenzymes $NADH_2$ and $NADPH_2$ could be regarded as equivalent, in a sense, to ATP. During the breakdown of biological molecules to simpler compounds oxidation occurs and ATP is synthesized. During the synthesis of larger molecules from small ones reduction with $NADH_2$ or $NADPH_2$ occurs and ATP is used. In any given synthesis either the oxidation of $NADH_2$ to NAD (or $NADPH_2$ to NADP) or the breakdown of ATP to ADP may predominate. We saw in the last chapter that the synthesis of triose phosphate from carbon dioxide required both $NADPH_2$ and ATP, whereas the synthesis of polysaccharide from triose phosphate required a further supply of ATP but no further reduction. In this chapter we shall find that the synthesis of fatty acid from acetyl coenzyme A again requires both $NADPH_2$ and ATP.

We pointed out at the end of the last chapter (p. 130) that carbohydrate could not be synthesized from fatty acids. Fatty acids, on the other hand, can be synthesized from any compound that gives rise to acetyl coenzyme A. One of the chief sources of acetyl coenzyme A is, of course, carbohydrate (p. 97). Thus fat can readily be synthesized from carbohydrate, which is every glutton's experience.

The key reaction in fatty-acid synthesis is a rather unexpected one, the carboxylation of acetyl coenzyme A to malonyl coenzyme A. This reaction requires ATP and is catalysed by the enzyme acetyl coenzyme A carboxylase, which (like many carboxylases) requires the cofactor biotin (p. 28).

$$CH_3.CO.S.CoA + CO_2 + ATP \rightleftharpoons$$
$$COOH.CH_2.CO.S.CoA + ADP + P_i.$$

acetyl coenzyme A   malonyl coenzyme A   (1)

Because ATP is used up, the equilibrium constant of this reaction overwhelmingly favours formation of malonyl coenzyme A. We saw in the last chapter that the biosynthetic sequences leading to triose phosphate and to polysaccharide contain steps that are effectively irreversible; here too the biosynthetic sequence leading to fatty acid contains an 'irreversible' reaction.

The malonyl group from malonyl coenzyme A is now bound to a small protein called the 'acyl carrier protein' (ACP). The linkage between the malonyl group and the ACP is a thiol ester, similar to the linkage between the malonyl group and coenzyme A.

$$COOH.CH_2.CO.S.CoA + ACP.SH \rightleftharpoons$$
$$COOH.CH_2.CO.S\text{-}ACP + CoASH. \quad (2)$$

In an exactly similar reaction, the acetyl group from another molecule of acetyl coenzyme A is also transferred to an acyl carrier protein.

$$CH_3CO.S.CoA + ACP\text{-}SH \rightleftharpoons CH_3.CO.S\text{-}ACP + CoASH. \quad (3)$$

It is now possible for the two units to condense together. This reaction occurs by an attack of the carboxyl carbon of acetyl ACP (marked with an asterisk) on the methyl carbon of malonyl ACP (marked with a). Carbon dioxide is eliminated.

$$CH_3.C^*O.S\text{-}ACP + COOH.C^aH_2.CO.S\text{-}ACP \rightleftharpoons$$
$$ACP\text{-}SH + CH_3.C^*O.C^aH_2.CO.S\text{-}ACP + CO_2. \quad (4)$$

acetyl-ACP   malonyl-ACP   acetoacetyl-ACP

This carbon dioxide is that that was used in reaction (1) to synthesize malonyl coenzyme A from acetyl coenzyme A. In a sense reaction (4) is really a condensation of two acetyl groups, and the carbon dioxide that made one of the two a malonyl group has served a catalytic role in the condensation.

The acetoacetyl group formed in reaction (4) must now be reduced. The first step in this process is the formation of a —CHOH group from the carbonyl group.

$$CH_3.CO.CH_2.CO.S\text{-}ACP + NADPH_2 \rightleftharpoons$$
$$CH_3.CHOH.CH_2.CO.S\text{-}ACP + NADP. \quad (5)$$

Next, water is eliminated, and the double bond that results is again reduced (we mentioned this latter reaction on p. 72).

$$CH_3.CHOH.CH_2.CO.S\text{-}ACP \rightleftharpoons CH_3.CH{=}CH.CO.S\text{-}ACP + H_2O$$
$$(6)$$

$$CH_3.CH{=}CH.CO.S\text{-}ACP + NADPH_2 \rightleftharpoons$$
$$CH_3.CH_2.CH_2.CO.S\text{-}ACP + NADP. \quad (7)$$

In principle, these reactions are the reverse of the three reactions involved in the oxidation of fatty acyl coenzyme A to β-ketoacyl coenzyme A (p. 104). There are, however, important differences. In the first place, each of the intermediates is tightly bound to its acyl carrier protein, and the intermediate–ACP complex is itself tightly bound to a complex containing the various enzyme activities necessary for these reactions. Secondly, whereas NAD is used as the acceptor of one pair of hydrogen atoms in the oxidation of fatty acids (the other pair being passed direct to a flavoprotein), the synthesis of fatty acids requires $NADPH_2$. (We recall that the principal source of $NADPH_2$ is the direct oxidation of glucose-6-phosphate (p. 108).) So while in principle it would be possible to oxidize fatty acids by a reversal of reactions (5), (6) and (7), in practice the degradative sequence and the synthetic sequence are quite distinct. Their separate identity is reinforced by the 'irreversible' steps in each sequence – the reactions catalysed by the thiokinase (p. 104) and β-keto thiolase (p. 105) in degradation, and the carboxylation reaction (1) in synthesis.

The reactions so far described have led to the production of butyryl-ACP and thus the synthesis of a fatty acyl group two carbon atoms longer than the original acetyl group has been achieved. It is now possible for another condensation to occur, similar to that in reaction (4). The butyryl group now takes the place of the acetyl group in reaction (4), and condenses with another malonyl group; carbon dioxide is again eliminated.

$$CH_3 . CH_2 . CH_2 . CO . S\text{-}ACP + COOH . CH_2 . CO . S\text{-}ACP \rightleftharpoons$$
$$CH_3 . CH_2 . CH_2 . CO . CH_2 . CO . S\text{-}ACP + ACP\text{-}SH + CO_2 \quad (8)$$

and the β-oxo-hexanoyl-ACP that results is now reduced by a repetition of reactions (5), (6) and (7).

$$CH_3 . CH_2 . CH_2 . CO . CH_2 . CO . S\text{-}ACP$$
$$\downarrow NADPH_2$$
$$CH_3 . CH_2 . CH_2 . CHOH . CH_2 . CO . S\text{-}ACP$$
$$\downarrow$$
$$CH_3 . CH_2 . CH_2 . CH{=}CH . CO . S\text{-}ACP$$
$$\downarrow NADPH_2$$
$$CH_3 . CH_2 . CH_2 . CH_2 . CH_2 . CO . S\text{-}ACP$$
hexanoyl-ACP

In this way long-chain fatty acids are built up. The essentials of each cycle of reactions is condensation of an acyl group with a malonyl group (each attached to its acyl carrier protein), followed by reduction with two molecules of $NADPH_2$. We can summarize the synthesis of stearoyl-ACP as follows.

$$CH_3 . CO . S\text{-}ACP + 8COOH . CH_2 . CO . S\text{-}ACP + 16NADPH_2 \longrightarrow$$
$$CH_3 . (CH_2)_{16} . CO . S\text{-}ACP + 8CO_2 + 8ACP\text{-}SH + 16NADP + 8H_2O.$$

This formulation emphasizes three facts. First the complex of intermediates, enzymes and acyl carrier proteins is a tight one (compare the complexes we mentioned on p. 36); and the cycle of condensation and reduction continues in this complex, without the liberation of intermediates, until the long-chain fatty acyl group is completed. Secondly, all of the two-carbon units (except the first) that are condensed into fatty acids enter the sequence as malonyl groups. Thirdly, each malonyl group had to be synthesized in a reaction involving ATP and each has to be reduced by two molecules of $NADPH_2$. Another way of writing the overall reaction may make this last point even clearer.

$$9CH_3 . CO . S . CoA + 8CO_2 + 8ATP + 16NADPH_2 \longrightarrow$$
$$CH_3 . (CH_2)_{16}CO . S . CoA + 8CO_2 + 8ADP + 8P_i + 16NADP$$
$$+ 8CoASH + 8H_2O.$$

We must now consider how a triglyceride is synthesized. The starting compounds are fatty acyl coenzyme A and α-glycerophosphate. Fatty acyl coenzyme A is formed from fatty acyl-ACP by a reversal of a reaction similar to (2) or (3).

$$CH_3 . (CH_2)_n . CO . S\text{-}ACP + CoASH \rightleftharpoons$$
$$CH_3 . (CH_2)_n . CO . S . CoA + ACP.$$

α-Glycerophosphate can be formed by phosphorylation of glycerol (p. 94).

$$
\begin{array}{ll}
CH_2OH & CH_2OH \\
| & | \\
CHOH \ +ATP \rightleftharpoons CHOH \ +ADP \\
| & | \\
CH_2OH & CH_2O\circledP \\
\\
\text{glycerol} & \alpha\text{-glycerophosphate}
\end{array}
$$

or by reduction of dihydroxyacetone phosphate:

$$
\begin{array}{cc}
\mathrm{CH_2OH} & \mathrm{CH_2OH} \\
| & | \\
\mathrm{CO} \quad +\mathrm{NADH_2} \rightleftharpoons & \mathrm{CHOH} \quad +\mathrm{NAD} \\
| & | \\
\mathrm{CH_2O\textcircled{P}} & \mathrm{CH_2O\textcircled{P}} \\
\text{dihydroxyacetone phosphate} & \alpha\text{-glycerophosphate}
\end{array}
$$

Two molecules of fatty acyl coenzyme A can react with one molecule of α-glycerophosphate to give a compound known as phosphatidic acid.

$$
\begin{array}{cc}
\mathrm{CH_3.(CH_2)_n.CO.S.CoA} & \\
+ & \mathrm{CH_2O.CO.(CH_2)_n.CH_3} \\
\mathrm{CH_3(CH_2)_m.CO.S.CoA} & | \\
+ & \rightleftharpoons \quad \mathrm{CHO.CO.(CH_2)_m.CH_3} \\
\mathrm{CH_2OH} & | \\
| & \mathrm{CH_2O\textcircled{P}} \\
\mathrm{CHOH} & \text{phosphatidic acid} \\
| & \\
\mathrm{CH_2O\textcircled{P}} &
\end{array}
$$

Phosphatidic acid can now be hydrolysed to yield a diglyceride, and this will react with a third molecule of fatty acyl coenzyme A.

$$
\begin{array}{cc}
\mathrm{CH_2O.CO.(CH_2)_n.CH_3} & \mathrm{CH_2O.CO.(CH_2)_n.CH_3} \\
| & | \\
\mathrm{CHO.CO.(CH_2)_m.CH_3} \rightleftharpoons & \mathrm{CHO.CO.(CH_2)_m.CH_3} + \mathrm{P}_i \\
| & | \\
\mathrm{CH_2O\textcircled{P}} & \mathrm{CH_2OH} \\
& \text{diglyceride}
\end{array}
$$

$$
\begin{array}{cc}
\mathrm{CH_2O.CO.(CH_2)_n.CH_3} & \mathrm{CH_2O.CO.(CH_2)_n.CH_3} \\
| & | \\
\mathrm{CHO.CO.(CH_2)_m.CH_3} \rightleftharpoons & \mathrm{CHO.CO.(CH_2)_m.CH_3} + \mathrm{CoASH} \\
| & | \\
\mathrm{CH_2OH} & \mathrm{CH_2O.CO.(CH_2)_l.CH_3} \\
+ & \text{triglyceride} \\
\mathrm{CH_3(CH_2)_l.CO.SCoA} &
\end{array}
$$

The final product of this sequence is thus triglyceride.

17 Amino acid metabolism

In the chapters that described the main sources of ATP and reducing power we made no mention of the use of nitrogen-containing substances as substrates. By contrast in Chs. 3 to 6 we stressed that the nitrogen-containing macromolecules proteins and nucleic acids possess great specificity, and that proteins in particular play extremely diverse roles that depend on their specific structure. The chief biochemical importance of amino acids, then, is not so much that they can be broken down to provide ATP, but rather that they are the constituents of proteins; so that it is appropriate to concentrate in this chapter chiefly on the *synthesis* of amino acids. However, amino acids can, in fact, serve as minor sources of ATP, so we shall also touch briefly on this aspect of their metabolism.

The most important fact that we wish to stress about amino acid metabolism is that it does not comprise a completely separate set of reactions unconnected with the sequences that we have described in previous chapters. The metabolism of amino acids is closely linked with that of carbohydrates and fats. In particular there is one reaction that is central to amino acid metabolism, providing a link between amino acids and the Krebs cycle.

$$
\begin{array}{c}
\text{COOH} \\
| \\
\text{CO} \\
| \\
\text{CH}_2 + \text{NH}_3 + \text{NADH}_2 \ or \\
| \qquad\qquad\quad \text{NADPH}_2 \\
\text{CH}_2 \\
| \\
\text{COOH} \\
\text{α-oxoglutaric acid}
\end{array}
\rightleftharpoons
\begin{array}{c}
\text{COOH} \\
| \\
\text{CHNH}_2 \\
| \\
\text{CH}_2 + \text{H}_2\text{O} + \text{NAD} \ or \\
| \qquad\qquad\quad \text{NADP} \\
\text{CH}_2 \\
| \\
\text{COOH} \\
\text{glutamic acid}
\end{array}
\qquad (1)
$$

In this reaction α-oxoglutaric acid is reductively aminated to yield glutamic acid, the hydrogen atoms necessary for the reduction being provided by one of the reduced coenzymes. The equilibrium

[136]

constant of this reaction is not far from one, and the back reaction provides a means of oxidizing glutamic acid with the liberation of ammonia. The enzyme that catalyses the equilibrium is called glutamate dehydrogenase.

The great importance of this reaction is that in most organisms it is the sole means by which inorganic nitrogen can be fixed into a non-nitrogenous organic molecule to form an amino acid. We shall see (p. 155) that there is a separate reaction for the fixation of ammonia to form carbamyl phosphate and thus pyrimidines; and also (p. 155) a reaction for the fixation of another molecule of ammonia into glutamic acid to form glutamine and thus purines. But the *de novo* synthesis of amino acids depends almost entirely on the synthesis of glutamic acid from α-oxoglutaric acid, since the amino group, once introduced into glutamic acid, can become transferred to yield other amino acids.

This type of transfer reaction is called a transamination, and it can be represented as follows.

$$
\begin{array}{c}
\text{COOH} \\
| \\
\text{CHNH}_2 \\
| \\
\text{CH}_2 \\
| \\
\text{CH}_2 \\
| \\
\text{COOH}
\end{array}
+
\begin{array}{c}
\text{R} \\
| \\
\text{CO} \\
| \\
\text{COOH}
\end{array}
\rightleftharpoons
\begin{array}{c}
\text{COOH} \\
| \\
\text{CO} \\
| \\
\text{CH}_2 \\
| \\
\text{CH}_2 \\
| \\
\text{COOH}
\end{array}
+
\begin{array}{c}
\text{R} \\
| \\
\text{CHNH}_2 \\
| \\
\text{COOH}
\end{array}
\qquad (2)
$$

Its essential feature is an exchange of an α-amino group with an α-keto group, catalysed by an enzyme called a transaminase. Pyridoxal phosphate is required as coenzyme; this is formed from pyridoxine, which is a vitamin (see p. 27). Much the commonest transaminations (in some organisms the *only* transaminations) are those in which one of the reactants is glutamic acid. The other reactant may be any one of a number of α-keto acids; the specificity of the enzyme determines which α-keto acid receives the amino group from glutamic acid.

One of the two products of the transamination is, of course, the amino acid corresponding to the keto acid that participated in the reaction. The other product is α-oxoglutaric acid; and this can now once more be converted to glutamic acid by reaction (1). Thus

the couple glutamic acid–α-oxoglutaric acid can function cata-
lytically in the synthesis of an amino acid, indirectly fixing ammonia into the α-keto acid.

$$
\text{ammonia} \quad\begin{array}{c} \longrightarrow \quad \text{glutamic acid} \\[1em] \longleftarrow \quad \text{α-oxoglutaric acid} \end{array}\quad\begin{array}{c} \longleftarrow \quad R.CO.COOH \\[1em] \longrightarrow \quad R.CHNH_2.COOH \end{array}
\tag{3}
$$

and provided that the Krebs cycle is functioning normally (in other
words if there is adequate metabolism of carbohydrate, see p. 102)
the supply of α-oxoglutaric acid will be sufficient for this purpose.

So far we have concentrated on the synthesis of glutamic acid by
fixation of ammonia and the use of the amino group thus formed
in the synthesis of other amino acids. But the $\Delta G^{0'}$ of both reaction
(1) and reaction (2) is small; thus the couple glutamic acid–α-
oxoglutaric acid can equally well function in the deamination of
amino acids.

$$
\text{ammonia} \quad\begin{array}{c} \longleftarrow \quad \text{glutamic acid} \\[1em] \longrightarrow \quad \text{α-oxoglutaric acid} \end{array}\quad\begin{array}{c} \longrightarrow \quad R.CO.COOH \\[1em] \longleftarrow \quad R.CHNH_2.COOH \end{array}
\tag{4}
$$

These reactions indirectly remove ammonia from an amino acid,
leaving the corresponding α-keto acid which can be further
metabolized (see below).

A further use of the glutamic acid–α-oxoglutaric acid couple is
in the transfer of an amino group from one amino acid to another
without the intervention of free ammonia. This reaction is useful
in circumstances when one amino acid is relatively abundant and
some others in short supply, and it requires the use of two
transaminases.

$$
\begin{array}{c} R'.CO.COOH \\[1em] R'.CHNH_2.COOH \end{array}\quad\begin{array}{c} \longleftarrow \quad \text{glutamic acid} \\[1em] \longrightarrow \quad \text{α-oxoglutaric acid} \end{array}\quad\begin{array}{c} \longleftarrow \quad R.CO.COOH \\[1em] \longrightarrow \quad R.CHNH_2.COOH \end{array}
\tag{5}
$$

By the use of these various reactions, organisms can convert any
of the α-keto acids formed in the common metabolic reaction
sequences into the corresponding amino acids. Thus alanine can

be produced from the pyruvic acid that is an intermediate in the Embden–Meyerhof pathway,

$$CH_3.CO.COOH \qquad\qquad COOH.CHNH_2.CH_2.CH_2.COOH$$

$$CH_3.CHNH_2.COOH \qquad COOH.CO.CH_2.CH_2.COOH \qquad (6)$$

and aspartic acid can be produced from the oxaloacetic acid that is an intermediate in the Krebs cycle.

$$COOH.CO.CH_2.COOH \qquad\qquad COOH.CHNH_2.CH_2.CH_2.COOH$$

$$COOH.CHNH_2.CH_2.COOH \qquad COOH.CO.CH_2.CH_2.COOH \qquad (7)$$

Similarly an amino acid can be formed from a keto acid which, though not itself a common metabolic intermediate, can be readily made from a starting material that is on a normal metabolic pathway. An example is the synthesis of serine. 3-Phosphoglyceric acid (p. 92) is oxidized, then transaminated, and finally dephosphorylated by the specific enzyme phosphoserine phosphatase.

$$\begin{array}{llll} \text{NAD} & \text{NADH}_2 \\ \text{COOH} & \text{COOH} & \text{COOH} & \text{COOH} \\ | & | & | & | \\ \text{CHOH} & \text{CO} & \text{CHNH}_2 & \text{CHNH}_2 \\ | & | & | & | \\ \text{CH}_2\text{O}\,\textcircled{P} & \text{CH}_2\text{O}\,\textcircled{P} & \text{CH}_2\text{O}\,\textcircled{P} & \text{CH}_2\text{OH} \end{array} \qquad (8)$$

Many of the reaction sequences that lead to the formation of amino acids are rather complicated; but the transformations involved are generally similar to those that we have already outlined. A few include special reactions – for example the synthesis of the rings of phenylalanine and tyrosine, and of histidine – but we shall not consider these here. The point that we wish to establish is that there are two ways in which the synthesis of amino acids is connected to the metabolic sequences that we have previously described. First there is the provision of the amino group by transamination involving the glutamic acid–α-oxoglutaric acid couple. Secondly, as we have seen in reactions (6), (7) and (8), there is the provision of the carbon atoms that form the backbone of the molecule.

Two of the most important starting compounds from which the backbones of the amino acids are formed are aspartic acid and glutamic acid. Aspartic acid can give rise to methionine, lysine and threonine; glutamic acid can give rise to proline and arginine. (Note that this use of glutamic acid for providing the carbon atoms of proline and arginine is separate from its function of donating the amino group in transaminations.) Now aspartic acid is produced by transamination of oxaloacetic acid, and glutamic acid (either directly or by transamination) from α-oxoglutaric acid. These reactions, then, represent major uses of carbon atoms from intermediates in the Krebs cycle (see Fig. 9.3), and are part of the reason why the carboxylation reaction for replenishing oxaloacetic acid (p. 101) is so essential.

What we have said in the last few pages may imply that all amino acids can be readily synthesized from starting materials that are present among the intermediates of carbohydrate and fat metabolism. This is true only of some organisms. Most animals, for example, are unable to synthesize all twenty of the amino acids; as a rough rule, most animals seem to be able to synthesize about half the total number. The remaining amino acids have to be provided in the diet (see p. 27), and these are known as essential amino acids (the precise list of essential amino acids varies slightly from one animal species to another). As these are needed in substrate quantities for the synthesis of protein they are readily distinguishable from the vitamins that are needed in catalytic quantities to act as cofactors (see p. 27). An amino acid is essential simply because the animal is unable to carry out one of the enzymic steps needed for its synthesis – generally one of the enzymic steps needed for the synthesis of the carbon-atom backbone. Thus, for example, although in plants and many micro-organisms aspartic acid can (as we have just mentioned) be converted to methionine, lysine and threonine, in most higher animals all of these amino acids are essential.

Reaction (5) above showed how one amino acid could donate its amino group, via glutamic acid, to form another amino acid. This conversion, however, is possible only if the appropriate keto acid

(R.CO.COOH in reaction (5)) is available. A required amino acid cannot be synthesized if the corresponding α-keto acid is not available, however much of the donor amino acid (R′.CHNH$_2$.COOH) is present. For this reason the nutritional value of a diet to an animal is not dependent merely on the quantity of protein it contains but also on the amino acid composition of the protein. An animal can make use of proteins only to the extent that they satisfy its requirement of essential amino acids. If the proteins are lacking in an essential amino acid then the other amino acids present are of no use in satisfying nutritional requirements, and their nitrogen will be split off as ammonia (see reaction (4)). Since proteins in the diet never contain amino acids in precisely the proportions that the animal requires, there is always some wastage of amino acids and a corresponding release of their deamination products, α-keto acids.

This release of α-keto acids provides substrates that act (as we remarked at the beginning of this chapter) as minor sources of fuel. We do not intend to describe in detail the way in which each of the twenty amino acids can be broken down to provide ATP, but it is possible to outline certain general principles.

Since the deamination product of an amino acid is an α-keto acid, we can expect that this will be treated in the way that is usual for breakdown of α-keto acids. We have seen (p. 97) that the α-keto acid that is an important product of metabolism of carbohydrates, pyruvic acid, is broken down by oxidative decarboxylation and yields acetyl coenzyme A. In an analogous way the α-keto acid that is an intermediate in the Krebs cycle, α-oxoglutaric acid, is broken down by oxidative decarboxylation and yields succinyl coenzyme A (p. 99). Although the α-keto acids derived from amino acids do not invariably undergo oxidative decarboxylation they do so quite generally. Again, as these two examples show, the product of oxidative decarboxylation of an α-keto acid is an acyl coenzyme A; and we may reasonably expect that this will be metabolized in the way that we described for acyl coenzyme A derivatives formed from fatty acids (Ch. 12).

The terminal product of this kind of metabolic process is normally either acetyl coenzyme A on the one hand, or a Krebs cycle intermediate or pyruvic acid on the other hand. [Pyruvic acid is grouped with the Krebs cycle intermediates because by carboxylation it can give rise to oxaloacetic acid (p. 101), whereas acetyl coenzyme A cannot be converted to a Krebs cycle intermediate (p. 130).] The distinction between amino acids that are broken down to acetyl coenzyme A, and those that are broken down to a Krebs cycle intermediate or pyruvic acid, was the basis of an old classification of the amino acids as 'ketogenic' or 'glucogenic'. This classification relied on the fact that in a diabetic animal acetyl coenzyme A exacerbates ketosis by increasing the concentration of acetoacetyl coenzyme A (see p. 107), whereas Krebs cycle intermediates can be converted to glucose (see pp. 125–8). The distinction is used less than it was, since we can now trace the entire metabolic sequence of the degradation of amino acids rather than being aware of no more than their end products.

Paradoxically, it is an amino acid that does not fit neatly into either of these categories, namely isoleucine, that provides the best illustration of amino acid breakdown. Isoleucine is both ketogenic and glucogenic, as we can see by the following sequence.

$$\begin{array}{c} CH_3 \\ {}^{}\diagdown \\ CH_3.CH_2{}^{\diagup} \end{array} CH.CHNH_2.COOH \rightleftharpoons \begin{array}{c} CH_3 \\ {}^{}\diagdown \\ CH_3.CH_2{}^{\diagup} \end{array} CH.CO.COOH$$

First, transamination with α-oxoglutaric acid gives an α-keto acid, and this, by reactions analogous to those for pyruvic acid (p. 97), is oxidatively decarboxylated.

$$\begin{array}{c} CH_3 \\ {}^{}\diagdown \\ CH_3.CH_2{}^{\diagup} \end{array} CH.CO.COOH \rightleftharpoons \begin{array}{c} CH_3 \\ {}^{}\diagdown \\ CH_3.CH_2{}^{\diagup} \end{array} CH.CO.S.CoA$$

The resulting acetyl coenzyme A enters the sequence for degradation of fatty acyl coenzyme A by β-oxidation (Ch. 12) and, by the β-keto thiolase reaction (p. 105), the molecule is split to yield acetyl coenzyme A and propionyl coenzyme A.

$$\begin{array}{c} CH_3 \\ {}^{}\diagdown \\ CH_3.CH_2{}^{\diagup} \end{array} CH.CO.S.CoA \rightleftharpoons \begin{array}{c} CH_3 \\ {}^{}\diagdown \\ CH_3.CH{}^{\diagup} \end{array} C.CO.S.CoA \rightleftharpoons$$

$$\begin{array}{c} CH_3 \\ {}^{}\diagdown \\ CH_3.CHOH{}^{\diagup} \end{array} CH.CO.S.CoA \xrightarrow[\qquad]{NAD \quad NADH_2} \begin{array}{c} CH_3 \\ {}^{}\diagdown \\ CH_3.CO{}^{\diagup} \end{array} CH.CO.S.CoA$$

$$\begin{matrix} CH_3 \\ \diagdown \\ \diagup \\ CH_3.CO \end{matrix} CH.CO.S.CoA \rightleftharpoons \begin{matrix} CH_3.CH_2.CO.S.CoA \\ + \\ CH_3.CO.S.CoA \end{matrix}$$

Now acetyl coenzyme A is the end product that would normally cause us to classify isoleucine as ketogenic; but on the other hand propionyl coenzyme A can be carboxylated and converted to succinyl coenzyme A, which is a Krebs cycle intermediate.

We have outlined this degradation sequence not because the breakdown of isoleucine is an important means of gaining ATP but because it illustrates four principles. In the first place, isoleucine, like most amino acids, can lose its amino group to α-oxoglutaric acid and thus form ammonia [see reaction (4)]. Secondly, once the amino group has been removed, the resulting product can no longer be regarded as belonging to a metabolic sequence that is peculiar to amino acids. Thirdly, the final products of the splitting of the molecule can enter the normal pathways of oxidation that we have already described; so to the extent that amino acids are used as fuel the Krebs cycle is their terminal pathway for catabolism. Finally, a part of the molecule can be converted to one of the Krebs cycle intermediates rather than to acetyl coenzyme A; this can be converted to oxaloacetic acid and thus, by the sequence described on pp. 125–9, to carbohydrate. In an analogous way succinyl coenzyme A can also be formed from valine, α-oxoglutaric acid can be formed from several amino acids (e.g. histidine, arginine and proline), pyruvic acid from several others (e.g. serine, alanine and cysteine), and oxaloacetic acid from aspartic acid; and all these can give rise to carbohydrate.

Interlude. We have now covered in some detail most of the pathways that we skimmed over in Ch. 9. (We have still to consider the synthesis of the purine and pyrimidine nucleotides, which we shall come to in Ch. 19.) This, then, might be a good moment for the reader to refer again to the figures in Ch. 9 which summarize the main pathways of intermediary metabolism, and to see how the principles that we outlined in that chapter have been worked out in the reaction sequences discussed in Chs. 10–17.

Activation of amino acids

At the beginning of this chapter we remarked that the chief bio-chemical importance of amino acids is the fact that they are constituents of proteins. We shall now start to see how proteins are in fact synthesized from amino acids. In Ch. 15 we showed how, in the synthesis of polysaccharide, glucose-1-phosphate has first to be converted to uridine diphosphate glucose before polymerization. We find that an analogous activation is necessary before amino acids can form peptide bonds; the reaction for an amino acid is the formation of the high energy compound aminoacyl-AMP.

$$R.CHNH_2.COOH + ATP \rightleftharpoons R.CHNH_2.CO-AMP + (P-P)_i.$$

The equilibrium constant of this reaction is near to 1; but as we have seen (p. 104) the fact that pyrophosphate is produced ensures that in practice the reverse reaction to synthesize ATP does not occur.

Aminoacyl-AMP contains a 'high energy' bond and is analogous to uridine diphosphate glucose (p. 129). One might therefore expect that aminoacyl-AMP would be capable of condensing to form a polyamino acid, splitting out AMP, in the same way as UDP-glucose condenses to form a polyglucose, splitting out UDP. In fact, such a reaction does not occur; and we must now consider why this is.

The reason is exactly that which we gave in Ch. 2. Glucose polymers are used for the storage of fuel in the form of carbohydrate, and they contain only one kind of monomeric residue. No question of arrangement of residues arises; even the number of residues in a polymer is not precisely determined. There is no specificity of structure and no need to ensure that one copy of a polymer molecule exactly resembles any other. Proteins, by contrast, are possessed of great specificity which allows them to perform all the functions that we outlined in Chs. 4 and 5. This specificity depends on an extremely precise ordering of the residues, and on an extreme reproducibility of the structure that ensures that every copy of the molecule is identical to every other.

Thus in the polymerization both of glucose and of amino acids, the residues must first be activated in the way that we have described. But in the synthesis of a glucose polymer no ordering of residues is required; so it is possible simply with the specificity of a single enzyme to ensure the formation of the correct linkage in the polymer (see p. 129). If proteins were random polymers of amino acids a similar condensation reaction would suffice; but since they are not, it is essential to find other means of ensuring that the precise ordering of amino acid residues is maintained in protein synthesis.

How is this precision of ordering brought about? We know that the exact structure of many proteins is characteristic not just of a single individual of a species, but of the species as a whole, and that it is maintained through the generations (for example, it can be deduced that pig insulin has precisely the same structure if isolated from a pig now as it had in an ancestral pig). So it seems that the ordering of amino acid residues in a protein is dependent on the genetic material of the organism, and, indeed, that it is one of the functions of the genetic material to determine the ordering. Therefore before we can discuss how the precision of structure of proteins is achieved, we must first describe the functions and the mode of synthesis of the genetic material.

Section III

MOLECULAR GENETICS AND PROTEIN SYNTHESIS

MOLECULAR GENETICS AND PROTEIN

18 The molecular basis of genetics

In Chs. 3–6 we outlined the ways in which structure and function are related in macromolecules. We did not describe in any detail one of the essential functions that macromolecules perform: the handling of *information*.

If one were entirely ignorant of the constitution of living matter at the molecular level one would hardly have guessed that the linear sequence of residues in a macromolecule could be used for the storage and transfer of information. The elaboration of molecules able to fulfil the role of information storage, and the development of mechanisms that bring about the transfer of information from one molecule to another, must rank as one of the great distinguishing features of living matter. It is of comparable importance with the development of macromolecules to overcome the limitations of the chemistry of free solution, with the extensive coupling between endergonic and exergonic reactions, and with the organization of metabolism into a large number of interlocking pathways.

It is for this reason that a whole section of this book has been devoted to informational macromolecules.

It is well known that DNA is chief among these macromolecules. In the rest of this chapter we will examine its role a little further and discuss its relation to genetics.

In many viruses, in all other micro-organisms and in all plants and animals, the genetic material is DNA. There is a wealth of experimental evidence that supports this statement, but we shall here mention only one kind of experiment. It is possible to extract and purify the DNA from a culture (the 'donor') of one particular species of bacteria and add it to another culture (the 'recipient'). Under certain conditions the recipient bacteria will take up the purified DNA and incorporate a piece of it in place of a piece of their own DNA. When this happens, the recipient bacteria may be found to have permanently acquired a characteristic of the donors – in other words, their genetic complement has been to that extent altered.

We remarked at the end of the last chapter that the genetic material appears to determine the structure of proteins, and the example that we gave was the control of the structure of insulin in the pig. If we wish to discover how the genetic material carries out this function, one obvious method is to see how *changes* in DNA are reflected in changes in the structure of proteins; but before we could find an altered insulin we might have to isolate the protein from thousands of separate pigs and in each case examine the structure of the molecule.

Another possible approach is to rely on the fact that enzymes are proteins of highly specific structure (see Ch. 5) and therefore that changes in their structure very frequently lead to loss of enzymic activity. So if we could find an organism that had lost a particular enzyme, we might be able to relate the loss to some change that had occurred in the DNA. However, with most organisms it is by no means easy to find individuals lacking particular enzymes; for if the enzyme is essential to metabolic activity we shall never find an individual without it, and conversely if the enzyme is inessential we shall not easily be able to recognize its loss.

We can overcome these difficulties by working with micro-organisms, because with micro-organisms we can alter conditions in such a way as to compensate for the loss of an enzyme. Suppose, for example, that a bacterium that is normally capable of producing serine by the pathway that we have described on p. 139 suffers a change in its DNA that results in a loss of the enzyme phospho-serine phosphatase. Serine is a constituent of all the proteins that the organism makes (see p. 17) and consequently the bacterium will be unable to grow without it. If, however, we include serine in the cultivation medium the bacterium will grow quite normally and we shall be able to study it. (We may say in this case that serine has become an essential amino acid for the bacterium. This example suggests that the reason why some amino acids are essential (p. 27) for certain species of animals is that the DNA of these animals has, in the course of evolution, become changed in such a way that the enzymes necessary for their synthesis have been lost.)

A change in the structure of DNA is called a mutation, and an organism that carries a mutation is called a mutant; by contrast the 'normal' organism, in which the mutation occurred, is called the wild type. (This definition is not a rigorous one, since organisms are described as wild type largely by convention, but in practice difficulty seldom arises.) Genetics has for many years made use of mutations without being able to define the precise change that occurred when the DNA underwent mutation, but recently it has become possible to correlate some mutations in DNA directly with changes in the structure of proteins.

The basic technique used in genetic studies is to make *crosses* between genetically different strains of a species – either between one mutant strain and another or between wild-type and mutant strains. (There are various means of making crosses and examining the progeny that result, but we shall not go into them here.) The value of genetic experiments is that their results allow one to make *genetic maps*. Just as the map of a railway line reveals the order of the stations and indicates the distances between them, so a genetic map reveals the order of the mutations along the DNA molecule and gives some idea of the distances between them. For technical reasons that need not concern us, it is particularly easy to make very detailed maps of the DNA in bacteria as compared with higher organisms. We shall now mention some of the results that have emerged from bacterial genetics; almost all of them are derived from studies of the two closely related species *Escherichia coli* and *Salmonella typhimurium*.

The first point is that if several, independently isolated strains are found that are all deficient in a single enzyme, the mutations almost always map close together. To continue with the example that we gave previously, it may be possible to isolate several dozen separate strains of *E. coli* that are deficient in phosphoserine phosphatase; and if these mutations are mapped it is extremely likely that they will all be found to lie very close to one another in a small segment of DNA. This fact suggests that that segment of DNA is responsible for determining the structure of phosphoserine phosphatase, and that mutations in any of a number of places in that segment lead to loss of the phosphatase. We can speak of such a segment of DNA as a gene – in this particular example the gene for phosphoserine phosphatase.

Now mutations can be of several kinds. The simplest kind of mutation is one in which a base in the polydeoxyribonucleotide chain (see p. 50) is replaced by another base – for example adenine may be replaced by guanine (since the bases in DNA are complementary (see p. 51), it would be more accurate to say that the base pair adenine–thymine is replaced by the base pair guanine–cytosine). This kind of mutation is called base substitution. Another kind of mutation involves the loss of a whole stretch of nucleotides (perhaps even many hundred nucleotides) from the two strands of DNA; this is called deletion. Yet another kind is the insertion of a stretch of nucleotides that really belong elsewhere into the middle of a gene; this is called insertion.

We do not know in detail how these changes in the structure of DNA occur. They happen at a low rate 'spontaneously', i.e. without experimental intervention, but the rate of their occurrence can be greatly increased by treating bacteria either with certain chemicals or with some kinds of radiation. The mode of action of some of these experimental treatments is known, and we can imagine that the spontaneous mutations occur in analogous ways – e.g. because all living organisms are subject to some background radiation.

In mutants deficient in a particular enzyme that have arisen through base substitution it is often possible to find in the cell a protein that is extremely similar to the lost enzyme. Sometimes it has been possible to prove that this new protein is in fact identical to the lost enzyme except for a change in a single amino acid residue. Moreover, in some cases in which a whole set of mutations has been discovered in a single gene each mutant has been found to contain a distinct protein that is derived, by a change in a single amino acid residue, from the original enzyme. If now these mutants are used in genetic crosses it is possible to map the position of their mutations within the single gene; and from studies of the mutant proteins it is possible to construct a map of amino acid changes within the single protein. These two maps correspond extremely well. The result shows that the *linear sequence* of the DNA is precisely related to the *linear sequence* of the protein, in other words that the ordering of bases in the DNA precisely corresponds to (and therefore must determine) the ordering of amino acids in the protein. In Ch. 20 we shall see how this determining of the order of amino acids actually occurs.

Meanwhile we shall mention one further result of genetic studies with bacteria. We have already seen that mutations that lead to the loss of a single enzyme generally lie very close together. In bacteria, genes that determine the enzymes of one *pathway* often lie close together too. For example, there are three enzymes that are specifically needed for the fermentation of galactose and are not needed for other fermentations. The structures of these three enzymes are determined by three genes, and these are contiguous on the DNA of *E. coli*. A similar clustering of genes is often found for biosynthetic sequences. A striking example is the pathway of synthesis of histidine, which involves ten specific enzymes. The genes that determine the structure of these form an uninterrupted stretch of bacterial DNA. Clustering of this sort is not found universally even in bacteria (for instance the genes that determine the structures of the enzymes that synthesize arginine are scattered around the DNA of *E. coli*), but where it does occur it seems to be important in the control of enzyme synthesis (see Ch. 21).

In this chapter we have discussed the techniques and results of genetics almost entirely in terms of bacteria, but we do not wish to give the impression that there has not been extremely important and valuable work with other micro-organisms (especially fungi) and with higher organisms. A great deal is known, for example, about human genetics from the study of mutations in such proteins as haemoglobin. However, the techniques of genetics in higher organisms are a great deal more difficult than in bacteria. In man, for example, it is not socially acceptable for a scientist to make experimental crosses between genetically interesting individuals. Even in animals like mice the results of a cross take weeks or months to emerge, instead of days. Bacterial genetics, on the other hand, has progressed extremely rapidly since its beginnings only 25 years ago, and it is this progress that has led to the establishment of most of the principles we have described.

19 Synthesis of DNA and RNA

The genetic material of a cell has two essential functions. One is to control the activities of the cell by specifying the structure of its components, especially proteins; we have seen in cursory outline in the last chapter, and shall examine in greater detail in the next chapter, how the linear sequence of nucleotides in DNA is related to the linear sequence of amino acids in proteins. The second function of the genetic material is to provide for its own exact replication, so that (barring the accident of mutation) each of the two daughter cells formed on division of the parental cell contains a precise copy of the genetic material of its parent. In this chapter we shall describe how the DNA replicates, in other words how DNA is synthesized in the cell and how the synthesis is controlled in such a way that the structure of the new genetic material is exactly similar to that of the old. We shall also describe how RNA is synthesized, because that process has much in common with the synthesis of DNA, and because a description of RNA synthesis is necessary for a consideration (in the next chapter) of protein synthesis.

Both DNA and RNA are synthesized from nucleoside tri-phosphates – deoxyadenosine triphosphate, deoxyguanosine tri-phosphate, deoxycytidine triphosphate and thymidine triphosphate are the precursors of DNA, and adenosine triphosphate, guanosine triphosphate, cytidine triphosphate and uridine triphosphate are the precursors of RNA. These nucleoside triphosphates are simply the nucleoside monophosphates that we have already mentioned as the constituents of DNA and RNA (p. 51), each carrying two extra phosphate groups. ATP is, of course, a familiar example of a nucleoside triphosphate, and the others are analogous in structure. As we have seen previously (p. 50), ATP and deoxyATP contain adenine, and GTP and deoxyGTP contain guanine; both adenine and guanine are purines. CTP and deoxyCTP contain cytosine, UTP contains uracil and TTP contains thymine; cytosine, uracil and thymine are all pyrimidines. Most organisms are able to synthesize both purines and pyrimidines, and we shall briefly outline their pathways of synthesis, and the formation of the nucleoside triphosphates, before showing how polymerization of DNA and RNA occurs.

The synthesis of nucleotides

The pathways of synthesis of purines and pyrimidines are rather dissimilar, but one thing that they have in common is that each requires a separate reaction for the fixation of ammonia. We have already seen (p. 136) that one extremely important reaction for the fixation of ammonia is the synthesis of glutamic acid, and that this provides for the formation of the amino group in other amino acids. Now glutamic acid can in fact receive another molecule of ammonia in a reaction that forms glutamine:

$$
\begin{array}{ccc}
\text{COOH} & & \text{CONH}_2 \\
| & & | \\
(\text{CH}_2)_2 + \text{NH}_3 + \text{ATP} \rightleftharpoons & (\text{CH}_2)_2 + \text{ADP} + \text{P}_i \\
| & & | \\
\text{CHNH}_2 & & \text{CHNH}_2 \\
| & & | \\
\text{COOH} & & \text{COOH} \\
\text{glutamic acid} & & \text{glutamine}
\end{array}
$$

and glutamine, as we shall see below, is used in the synthesis of purines. The reaction for the fixation of ammonia that is involved in the synthesis of pyrimidines is the formation of carbamyl phosphate.

$$CO_2 + NH_3 + 2ATP \rightleftharpoons NH_2.CO.O\textcircled{P} + 2ADP + P_i.$$

Notice that both syntheses use up ATP; the fixation of ammonia into α-oxoglutaric acid that we described on p. 136 does not use ATP but instead uses a reduced coenzyme which can be regarded as equivalent (Ch. 7).

[We must digress to point out that both glutamine and carbamyl phosphate have important uses other than in the synthesis of purines and pyrimidines. Glutamine is a constituent of proteins, and it can also provide a nitrogen atom in the synthesis of the ring-containing amino acids histidine and tryptophan (and various other compounds). Carbamyl phosphate is essential for the synthesis of urea, which is the compound that many animals (including ourselves) form in order to excrete surplus nitrogen.]

The pathway of purine synthesis is complicated and we shall not consider it in any detail. Perhaps its most interesting feature is that the purine rings are built up, piece by piece, on a ribose-5-phosphate foundation. The effect of this is that, when the purine rings are completed, the molecule is already in the form of a nucleotide – actually the molecule inosinic acid (IMP), from which AMP and GMP can be easily formed. We recall that ribose-5-phosphate is formed by the oxidation of glucose-6-phosphate followed by isomerization (pp. 108–10), and in discussing the importance of this reaction we pointed out that ribose is an essential constituent of nucleotides. In the first reaction of purine biosynthesis, ribose-5-phosphate receives a pyrophosphate group from ATP to yield 5-phosphoribosyl pyrophosphate.

ribose-5-phosphate   5-phosphoribosyl pyrophosphate

5-phosphoribosyl   glutamine   5-phosphoribosylamine   glutamic acid
pyrophosphate

This molecule then reacts with glutamine to give 5-phosphoribosylamine. The nitrogen atom that is now attached to the 1-position of the ribose-5-phosphate remains in this position throughout the synthesis of the purine rings. During the synthesis several molecules react successively with the 5-phosphoribosylamine – including another molecule of glutamine which provides one of the other nitrogen atoms of the purine. The final products are AMP and GMP, whose structures are now given again.

AMP   GMP

GMP can receive another phosphate group from ATP in a kinase-catalysed reaction:

$$GMP + ATP \rightleftharpoons GDP + ADP.$$

The nucleoside *di*phosphates can be converted to deoxynucleoside diphosphates in a reaction that involves reduction with $NADPH_2$. GDP, deoxyADP and deoxyGDP can all receive a third phosphate group from ATP in another kinase-catalysed reaction. This completes the synthesis of the four purine nucleoside triphosphates which, as we shall see below, are the substrates for the polymerase enzymes that catalyse the synthesis of DNA and RNA.

The synthesis of pyrimidines, by contrast, does not take place on a ribose-5-phosphate molecule. Instead, the pyrimidine ring is formed first and then converted to a nucleotide. The first reaction in the formation of pyrimidines is the synthesis of carbamyl aspartic acid from carbamyl phosphate and aspartic acid.

carbamyl phosphate   aspartic acid   carbamyl aspartic acid

This reaction is catalysed by the enzyme aspartate transcarbamyl-ase, which has been extensively studied as an example of an enzyme that is subject to 'end-product inhibition'. We shall discuss this phenomenon in some detail in Ch. 21. The use of aspartic acid [which derives from oxaloacetic acid (p. 139)] for the synthesis of pyrimidines is yet another example of the way in which Krebs cycle intermediates provide starting material for synthetic pathways (see p. 101 and compare p. 140, where we mentioned the use of aspartic acid for the synthesis of other amino acids).

Carbamyl aspartic acid is cyclized, and the product is oxidized. The resulting pyrimidine is called orotic acid.

It is at this stage that ribose-5-phosphate is attached. The reactant, once again, is 5-phosphoribosyl pyrophosphate.

The resulting orotidine-5′-phosphate is decarboxylated to yield uridine-5′-phosphate (uridylic acid or UMP).

UMP

By two successive reactions with ATP, UMP can be converted to UTP; and this, by reaction with ammonia, can give rise to CTP.

UTP   CTP

CDP, which derives from CTP, can be reduced to deoxyCDP, in a reaction with $NADPH_2$ similar to that involved in the formation of deoxyADP and deoxyGDP (p. 157). DeoxyCDP leads to deoxyCTP by phosphorylation with ATP; alternatively, by a more complicated series of reactions, deoxyCDP can give rise to TTP.†
This rather sketchy description accounts for the formation of deoxy-ATP, deoxyGTP, deoxyCTP and TTP, which are the building blocks from which DNA is synthesized, and we must now consider the mechanism of this synthesis. We have also accounted for the synthesis of ATP, GTP, CTP and UTP; later we shall see (p. 163) how these nucleoside triphosphates are used in the synthesis of RNA.

Synthesis of DNA and RNA

We have previously referred in some detail (p. 50) to the hydrogen bonding properties of the purine and pyrimidine bases. Let us now

† Thymidylic acid derivatives are, of course, normally deoxy-compounds since thymine is seldom found in RNA.

suppose that we wish to replicate a molecule of DNA, which we will call the parental molecule, and that we have a supply of all the deoxynucleotides. If we imagine the two strands of the parental molecule separating, we can see that each of the strands could act as a template which, owing to the specificity of hydrogen bonding between the bases, would ensure that a *complementary* sequence of nucleotides would become aligned opposite to itself. If these nucleotides now polymerized to form new strands of polynucleotides (daughter strands), we would have two double-stranded molecules of DNA. In each molecule the daughter strand would have the same sequence as the *opposite* parental strand (i.e. the parental strand in the other molecule). Consequently each molecule would consist of one parental strand and one daughter strand (that is, it would be half-old and half-new), and also each molecule would be identical in sequence to the original parental molecule.

An example will help to make this clear. If our parental double-stranded molecule has the sequence:

....A-A-C-T-G-G-G-G-T-T-C-C-A-T-G
....T-T-G-A-C-C-C-C-A-A-G-G-T-A-C

then the strands would separate to give:

....A-A-C-T-G-G-G-G-T-T-C-C-A-T-G

and   ....T-T-G-A-C-C-C-C-A-A-G-G-T-A-C.

Nucleotides would become aligned opposite these single strands in the following way (we use bold type to indicate the new nucleotides):

....A-A-C-T-G-G-G-G-T-T-C-C-A-T-G
....**T T G A C C C C A A G G T A C**

and   ....**A A C T G G G G T T C C A T G**
....T-T-G-A-C-C-C-C-A-A-G-G-T-A-C

and these nucleotides would polymerize:

....A-A-C-T-G-G-G-G-T-T-C-C-A-T-G
....**T-T-G-A-C-C-C-C-A-A-G-G-T-A-C**

and   ....**A-A-C-T-G-G-G-G-T-T-C-C-A-T-G**
....T-T-G-A-C-C-C-C-A-A-G-G-T-A-C.

These two DNA molecules are identical to one another, and they are also identical to the parental molecule. When the cell divides each daughter cell will receive one of the two molecules and will thus have exactly the same DNA as the parental cell.

This, then, in outline is how DNA is replicated. But the description so far has referred only to the way in which specific base-pairing is used to make two exact copies of the original DNA molecule. We have now to see how the replication occurs in terms of nucleotides rather than bases, and in particular to show how the polymerization reaction involves the deoxyribonucleoside triphosphates whose synthesis we have already described.

We have previously pointed out (p. 5) that if we consider carefully a single strand of polydeoxyribonucleotide we see that the two ends of the molecule are not identical. This difference between the ends is not a question of which bases are present, but a property of the sugar-phosphate backbone.

In this figure, we have drawn a tri-deoxyribonucleotide for the sake of simplicity, whereas in fact the strands of deoxyribonucleotides in DNA are extremely long. However, this miniature strand of DNA is sufficient to make clear the difference between the ends of the molecule. At the top we have a deoxyribose that has a free hydroxyl group on the 5' carbon atom. At the bottom we have a deoxyribose that has a free hydroxyl group on the 3' end of the molecule. (In the middle we have a deoxyribose that has neither a free 5' hydroxyl nor a free 3' hydroxyl group; and in a real strand of DNA there would, of course, be thousands or even millions of residues in this state.)

The route of synthesis of deoxyribonucleoside triphosphates that we described earlier in this chapter leads to deoxynucleosides with three phosphate groups attached to the 5' carbon atom of deoxyribose (see pp. 157 and 159). In the synthesis of DNA a molecule of this kind reacts with the free 3' hydroxyl end of a strand of polydeoxyribonucleotide.

This reaction produces inorganic pyrophosphate and therefore, for the reason that we discussed on p. 104, cannot be used in reverse for the degradation of DNA. Notice that, after a single deoxyribonucleotide has been added to the chain, a new 3′ hydroxyl end is left to which the next deoxyribonucleoside 5′ phosphate can attach.

In this way the new polydeoxyribonucleotide strand grows in the 5′ → 3′ direction. There are some unsolved problems about the nature of the enzyme (or enzymes) involved in the reaction, but we shall not go into them here. The important fact that we wish to stress is that the deoxyribonucleoside 5′ phosphate that is inserted into the growing (daughter) strand is selected by hydrogen bonding with the opposite nucleotide in the parental strand. This mechanism provides for the exact replication of DNA, which is essential if it is to act as genetic material.

We can now turn to the synthesis of RNA, which is analogous in many respects with the synthesis of DNA. An enzyme called RNA polymerase is known which, in the presence of DNA, catalyses the synthesis of ribonucleotide polymers. This enzyme employs nucleoside $5'$ triphosphates (products of the synthetic pathways we have described on pp. 155–9) and splits out one pyrophosphate for every nucleotide inserted into the polymer; it requires the presence of all four nucleoside triphosphates, and it synthesizes an RNA molecule in the $5' \rightarrow 3'$ direction.

The synthesis of RNA by this enzyme is dependent on the presence of DNA, which acts as a template that the enzyme copies, or 'transcribes'. This transcription involves pairing between the bases of the incoming ribonucleotides that are to form the RNA molecule and the bases of the deoxyribonucleotides in the template DNA. The RNA polymerase, however, transcribes only one strand of DNA and forms a single-stranded molecule of RNA that is similar in sequence (except that uracil is substituted for thymine) to the *opposite* strand of the DNA. Once again we use bold type to indicate the new nucleotides, this time those that are to form the RNA:

> ...**U-U-G-A-C-C-C-C-A-A-G-G-U-A-C**
> ...A-A-C-T-G-G-G-G-T-T-C-C-A-T-G
> ...T-T-G-A-C-C-C-C-A-A-G-G-T-A-C.

In this way it makes available an exact RNA transcript of the DNA. More accurately we should say that the RNA polymerase can make available an exact RNA transcript of any desired length of the DNA, since RNA molecules are far smaller than the DNA (see p. 54).

The great importance of this synthesis of RNA is that it enables protein synthesis to take place. In the next chapter we shall discuss in detail the role of RNA in the synthesis of proteins.

20 Synthesis of proteins

We outlined in Ch. 18 some of the evidence that leads to the conclusion that the linear sequence of the bases in DNA determines the linear sequence of amino acids in proteins. But in giving an account of RNA synthesis in Ch. 19 we implied that it is the RNA transcribed from DNA, rather than the DNA itself, that is actually involved in protein synthesis.

Most of the RNA that cells contain is present in small particles called *ribosomes* (see p. 54). Ribosomes consist of RNA and protein in the rough ratio of 60:40; bacterial ribosomes have a molecular weight of about 2.6 million, and ribosomes from higher cells a molecular weight of about 4–5 million. One interesting feature of ribosomes is that they seem always to consist of two unequal subunits: in bacteria (the ribosomes of which have been studied in most detail) these have molecular weights of about 1.8 million and 0.8 million. It is conventional to give ribosomes, and their subunits, names that refer to their rate of sedimentation in a centrifugal field. Thus in bacteria the bigger subunit is called the '50S particle' and the smaller the '30S particle', and the whole ribosome is called the '70S ribosome'. The corresponding names for the ribosomes of higher cells are '60S particle', '40S particle' and '80S ribosome'. The 30S particle of bacteria contains a single molecule of RNA (see p. 54) and about twenty molecules of protein (probably representing just one copy of each of twenty different proteins). The 50S particle contains two different molecules of RNA and about thirty-five different molecules of protein.

The reason why we have given this description of ribosomes is that they represent the site of synthesis of proteins; we shall see later just how it is that they function. What we must now consider is whether, as one might expect, it is the RNA of the ribosome that is involved in specifying protein structure. Ribosomal RNA, like all RNA, is synthesized by the RNA polymerase which transcribes the base sequence of DNA into a similar base sequence of RNA (see p. 163). Now we have seen (p. 153) that the sequence of bases in DNA ultimately determines the sequence of amino acids in protein. So it would seem reasonable to imagine that the ribosomal RNA acts as an intermediate in protein synthesis and that the structure of the protein, determined at one remove by DNA, is *directly* determined by the ribosomal RNA.

This idea turns out not to be correct. It is found, for example, that bacteria can change their pattern of protein synthesis without making new ribosomal RNA, and there is reason to believe that all ribosomes in a single cell may contain identical molecules of RNA even though they are synthesizing many different proteins. In other words, the ribosome contains machinery for *assembling* proteins (e.g. for making peptide bonds) but not the instructions for determining the sequence of their amino acids. A ribosome receives such instructions from time to time, and, until they are superseded, will follow them in assembling a protein.

These instructions are provided to the ribosome in the form of another species of RNA, which is called messenger RNA. Messenger RNA, too, is synthesized by the transcribing activity of RNA polymerase (p. 163); but in this case the sequence of nucleotides that are transcribed determines specifically the sequence of amino acids in proteins, whereas the DNA sequence that is transcribed to make ribosomal RNA appears not to correspond to any particular protein. A consequence of this difference is that most of the DNA of an organism, which is occupied with determining the structures of the thousands of different proteins synthesized, is transcribed into messenger RNA: by contrast there is only a very short sequence of DNA that is transcribed into ribosomal RNA.

Earlier (p. 152) we used the word 'gene' to describe a length of DNA the sequence of which corresponds to the sequence of a protein. On p. 163 we showed how RNA polymerase was capable of making an RNA transcript of any desired length of DNA. We can now put all these ideas together and say that RNA polymerase can transcribe a gene to produce the *messenger* RNA that corresponds to that gene. (We shall see in the next chapter that sometimes a messenger RNA may correspond to several contiguous genes.) Now since the ribosome receives its instructions for determining the sequence of a protein in the form of messenger

RNA, a ribosome that receives a messenger RNA that corresponds to the gene for phosphoserine phosphatase will thereby be instructed to synthesize phosphoserine phosphatase. In an analogous way ribosomes will be instructed to synthesize any given protein by receiving the messenger RNA that has been transcribed from the corresponding gene. So long as that messenger RNA remains intact they will synthesize that particular protein, but if after some time the messenger RNA is degraded (see p. 177) they will be free to accept a fresh messenger RNA and synthesize a new protein. In a sense, therefore, the ribosome is the slave of whatever messenger RNA happens to be bound to it at any given time.

So far we have spoken loosely of the messenger RNA 'instructing' the ribosome how amino acids should be arranged to make a protein. In fact the only specificity that a molecule of messenger RNA can possess must lie in its sequence of bases, in just the same way as is true of DNA (p. 153). How then can the sequence of bases in messenger RNA determine the sequence of amino acids in a protein? In order to answer this question, we must look at the fate of the activated amino acids that we described on p. 144. We left the amino acids in a sort of limbo, attached to AMP; but actually aminoacyl-AMP is an unstable intermediate that has only a transitory existence. The amino acids are immediately transferred again from AMP; and the molecules that now accept them belong to a third species of RNA which is called transfer RNA or tRNA (p. 54). This reaction is catalysed by the same set of enzymes as formed the aminoacyl-AMP intermediates; each enzyme is highly specific for a particular amino acid. We can now write the two reactions catalysed by these enzymes (whereas on p. 144 we artificially interrupted the process after the first reaction).

$$R.CHNH_2.COOH + ATP \rightleftharpoons R.CHNH_2.CO.AMP + (P-P)_i$$
$$R.CHNH_2.CO.AMP + tRNA \rightleftharpoons R.CHNH_2.CO.tRNA + AMP.$$

As pyrophosphate is produced the equilibrium of the total reaction is strongly in favour of the amino acyl-transfer RNA (see p. 104).

Each molecule of transfer RNA is specific for a particular amino acid, and since the enzymic reaction too is specific for the amino acid it follows that each amino acid is matched to its own transfer RNA molecule. Once an amino acid has been bound to its specific transfer RNA, it is the specific structure of the *transfer RNA* (rather than of the amino acid) that enables the complex to be recognized.

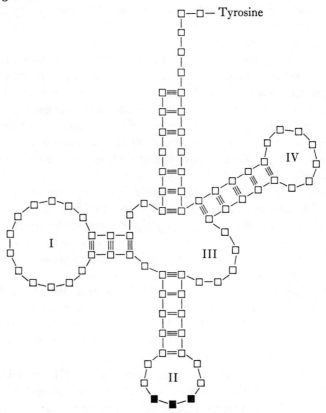

Fig. 20.1 Tyrosyl transfer RNA.

Transfer RNA molecules are exceptionally small among RNA species, having a molecular weight of about 25 000 and containing 70 to 80 nucleotides (see Fig. 20.1 and Table 6.1). The 3′ hydroxyl end of the molecule (see p. 5) always has the sequence – cytidine–cytidine–adenosine. The amino acid is bound to the 3′ hydroxyl group of the adenosine. Since this binding is common to all tRNA molecules, it cannot represent the specific feature by which each one is recognized.

In fact the recognition of tRNA molecules has two aspects. In the first place the molecule must interact with the specific enzyme that charges it with its amino acid: it is not certain how this recognition takes place. Secondly the tRNA molecule must be recognized during the translation of the messenger RNA by the ribosome; the tRNA can be shown to bind to the messenger only at the point where the translation process demands the incorporation of the particular amino acid that it carries. The recognition site for this step is known: it is a sequence of just three bases in the tRNA. In those tRNA molecules whose structure has been worked out it is found that this recognition site (marked in bold type in Figs. 20.1–5) is always in the second single-stranded region of the chain, counting from the 5′ end.

It might be convenient at this point to recapitulate, since we have now described the three chief reactants in the process that results in the synthesis of proteins. The *ribosome* is the site of synthesis of proteins, and although ribosomes contain RNA, this RNA does not determine the sequence of amino acids that are to be assembled. *Messenger RNA* carries the sequence of bases that dictates the sequence of amino acids. *Transfer RNA* carries the amino acids, and each molecule includes along its length a sequence of three bases which, since they are characteristic of that molecule, signals which amino acid is attached to its 3′ hydroxyl end.

We have seen (pp. 160 and 164) that hydrogen bonding between bases permits a sequence of bases in one polynucleotide chain to recognize a complementary sequence in another polynucleotide chain. We have previously mentioned this interaction in terms of the two strands of DNA, and in terms of pairing between DNA and RNA; but it is equally possible for a sequence of bases in one RNA molecule to recognize by hydrogen bonding a sequence of bases in another RNA molecule. So it should be possible for the sequence of three bases that is characteristic of each transfer RNA molecule to recognize a complementary sequence of three bases in messenger RNA. This idea suggests that sets of three bases in each molecule of messenger RNA might be used to specify which amino acids should be incorporated into the corresponding protein. Since there are four different bases in RNA, there are $4 \times 4 \times 4$

(= 64) possible combinations of three bases, and this is more than enough to account for all the amino acids that need to be specified.

It is now certain that this suggested method by which messenger RNA can dictate which amino acids must be assembled into the polypeptide chain is correct. Each of the amino acids corresponds to a sequence of three bases in messenger RNA, and to each of the possible three-base sequences an amino acid has been assigned (except for three of the 64, which are used to signal the end of the polypeptide chain). These three-base sequences are usually called *codons*, since the messenger RNA is often spoken of as 'coding for' a particular protein; the complementary sequence of three bases in each transfer RNA, which is characteristic of that molecule of transfer RNA and signals the nature of the amino acid that it is carrying, is called the *anti-codon*. Table 20.1 gives the assignment of codons to amino acids.

Table 20.1

UUU UUC } Phenylalanine	UCU UCC UCA UCG } Serine	UAU UAC } Tyrosine	UGU UGC } Cysteine
UUA UUG } Leucine		UAA　* UAG　*	UGA　* UGG　Tryptophan
CUU CUC CUA CUG } Leucine	CCU CCC CCA CCG } Proline	CAU CAC } Histidine CAA CAG } Glutamine	CGU CGC CGA CGG } Arginine
AUU AUC } Isoleucine AUA AUG　Methionine	ACU ACC ACA ACG } Threonine	AAU AAC } Asparagine AAA AAG } Lysine	AGU AGC } Serine AGA AGG } Arginine
GUU GUC GUA GUG } Valine	GCU GCC GCA GCG } Alanine	GAU　Aspartic GAC　acid GAA　Glutamic GAG　acid	GGU GGC GGA GGG } Glycine

* These codons signify termination of the polypeptide.

So far we have said little about the role of the ribosome. As we mentioned earlier, ribosomes are large and complicated organelles, and they are being intensively studied at present. The results of these studies have shown that ribosomes have many functions; but we can regard them for the present as fulfilling two tasks: first to bring the messenger RNA into juxtaposition with a succession of transfer RNA molecules, and secondly to hold the growing peptide chain. We can describe the mechanism of assembly of amino acids into protein by using as an example the translation of the small fragment of messenger RNA whose synthesis we described in the last chapter. This messenger RNA has the sequence ...UUGACCCCAAGGUAC... We have already seen that a sequence of three bases in a messenger RNA molecule specifies one amino acid, so that we can represent the fragmentary messenger RNA as ...UUG ACC CCA AGG UAC... By reference to Table 20.1 we can see that the corresponding peptide will be ...leucyl–threonyl–prolyl–arginyl–tyrosine... It is easiest to describe the process of elongation of the peptide chain, in other words to imagine the process in full swing, and we shall assume that the first codon (UUG) of our messenger RNA has just been translated – i.e. that the leucine has just been inserted into the polypeptide – so that the codon ACC is now waiting to be translated.

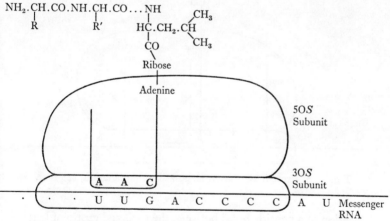

Fig. 20.2. Protein synthesis: stage 1. For explanation, see text.

We can represent this situation by the diagram given as Fig. 20.2. The messenger RNA is bound to the 30S subunit of the ribosome; the part of the messenger RNA that is to the left of the ribosome has already been translated, and the part that is to the right of the ribosome is going to be translated. The ribosome is holding a peptidyl transfer RNA; this represents the incomplete peptide that has already been synthesized, terminating in leucine and esterified to the leucine-specific transfer RNA. The last point will become clearer shortly; at this point what we wish to stress is that the peptide has grown from its free —NH_2 end so that it can be represented as:

$$\underset{|}{\overset{R}{}}\quad\underset{|}{\overset{R'}{}}\quad\underset{|}{\overset{R''}{}}\quad\underset{|}{\overset{R'''}{}}$$
$$NH_2.CH.CO.NH.CH.CO.NH.CH.CO \longrightarrow NH.CH.CO-tRNA$$

The ribosome will now accept and bind another aminoacyl tRNA – that which is specified by the codon ACC (threonyl tRNA) – and the resulting situation will be that represented in Fig. 20.3. In the next reaction the ribosome will catalyse the

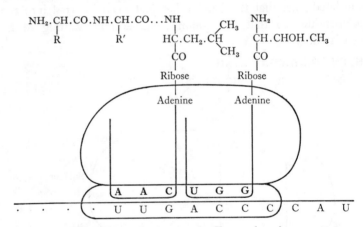

Fig. 20.3. Protein synthesis: stage 2. For explanation, see text.

formation of a peptide bond between the carboxyl group of leucine and the amino group of threonine. In other words leucine is transferred to threonine from its tRNA so that the leucine-specific tRNA is no longer esterified (Fig. 20.4). The ribosome will now move one codon along the messenger RNA, meanwhile holding the lengthened peptidyl tRNA (which now terminates in threonyl tRNA) opposite the threonine codon. (In this step the leucine-specific tRNA is ejected. It can be esterified again to leucine (see

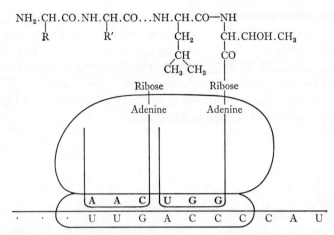

Fig. 20.4. Protein synthesis: stage 3. For explanation, see text.

p. 144) and thus take part once more in the reactions of protein synthesis.) Fig. 20.5, representing the position that we have now reached, is exactly analogous to Fig. 20.2. By repetitions of these reactions, the ribosome will catalyse the successive insertion of proline, arginine and tyrosine into the peptide.

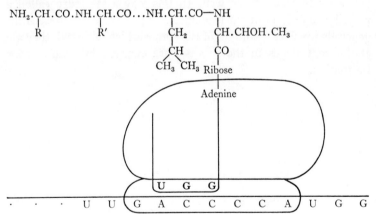

Fig. 20.5. Protein synthesis: stage 4. For explanation, see text.

There are two points that the above description omits but which should be mentioned. The first is that in describing *elongation* of the peptide chain we have avoided some rather special problems that are involved in beginning and ending the synthesis. The second is that the synthesis is an energetically costly process. Apart from the loss of two high energy bonds in the formation of aminoacyl-tRNA (p. 144), the insertion of each amino acid into the peptide chain involves the expenditure of at least two molecules of GTP during the steps illustrated in Figs. 20.2–5.

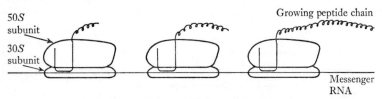

Fig. 20.6. Diagram of a polyribosome.

Thus ribosomes cause peptides to grow by moving along the messenger RNA and aligning aminoacyl-tRNA molecules opposite their corresponding codons. As each ribosome moves along the messenger RNA, it leaves free the beginning of the messenger RNA to which it attached; another ribosome can then bind to this point and start to translate as well. In this way a structure called a polyribosome is built up (see Fig. 20.6). Protein synthesis thus generally involves a number of ribosomes which, although they are at different stages in the process, are engaged in simultaneous translation.

21 Regulatory mechanisms in metabolism

In Section II of this book we outlined a number of metabolic reactions that cells are capable of carrying out, and the reader will have got the impression that this almost bewildering complexity demands careful organization and control. Similarly in the last two chapters we have explained the way in which the information that resides in the DNA can be expressed in terms of protein structure, and once again the reader may have found himself wondering how this complicated machinery is regulated. In this chapter we shall discuss some of the mechanisms by which cells control the rates of their enzymic reactions. In the first part of the chapter we shall consider the means by which the rate of enzyme *synthesis* is regulated (we may presume that control of the synthesis of non-enzymic proteins is similar); in the second part we shall consider the means by which enzymic *activity* is regulated.

Control of enzyme synthesis

We have seen (Ch. 19) that RNA polymerase is capable of making an RNA copy of DNA; and we have also seen that most of the DNA in the cell contains the genetic information that specifies the structure of proteins (Ch. 18). The RNA copy of this DNA is called messenger RNA (p. 166), and messenger RNA encodes in its sequence information that is translated by the ribosome into a sequence of amino acids (pp. 170–3).

Now although it is essential for the genetic material of a cell to contain all the information for specifying the structure of any protein that the cell is capable of making, at no time will the cell find a use for every one of the proteins that it can potentially make. We can illustrate this idea most easily by reference to bacteria. Wild-type *Escherichia coli* cells can synthesize all the amino acids from extremely simple starting materials – glucose, ammonia and a few inorganic salts. In order to synthesize (for example) histidine, they are obliged to make ten enzymes that are specific to the pathway of histidine biosynthesis (see p. 153). If, however, they find themselves in a medium in which histidine is present they can take up histidine from the medium, and in these circumstances they will no longer need to make the ten specific enzymes. A slightly different illustration is provided by the system for the utilization of lactose in *Escherichia coli*. Cells of this species are potentially capable of breaking down the disaccharide lactose, and they have the genetic information necessary to specify the structure of an enzyme (β-galactosidase) that hydrolyses lactose to the two hexoses, glucose and galactose, both of which can be metabolized by the Embden–Meyerhof pathway. However, if they are growing in glucose rather than lactose they will not need to make β-galactosidase since they have a direct source of hexose.

We have just remarked that in the presence of histidine there is no need for *E. coli* to make the enzymes of the histidine biosynthetic pathway, and that in the presence of glucose and the absence of lactose there is no need for them to make β-galactosidase. To say that there is no need, however, is to understate the case. We have seen in the last two chapters that the processes of messenger RNA synthesis and protein synthesis are complicated and energetically costly. To make proteins that are not needed, therefore, is not merely neutral in effect but disadvantageous. So there is good reason for cells to have a mechanism that actually *ensures* that superfluous proteins are not made.

The control of protein synthesis is a large subject, and we know a good deal about only a small fraction of it. We have some very clear information about the control of protein synthesis in bacteria, and much less information about the situation in higher cells. Even in bacteria, our knowledge is largely restricted to the two species *Salmonella typhimurium* and *E. coli*: and it is no accident that these are the two species the genetics of which have been most closely studied.

Before we describe in a little detail the system for regulating protein synthesis of which we have the most detailed understanding, it will be useful to review some of the possible means of control that organisms might, in principle, use. They can be divided into two classes – control at the level of messenger RNA synthesis and control at the level of the ribosome.

In our description (Ch. 20) of the mechanism of protein synthesis we tacitly assumed that the ribosome would translate without discrimination any messenger RNA that was present in the cell. If this assumption is correct, then the rate of synthesis of any particular protein will depend only on the concentration of its corresponding messenger RNA; the concentration of any messenger RNA will in turn depend on the rate of its synthesis and the rate of its degradation. On the other hand, it is possible that even if several kinds of messenger RNA are present in a cell at the same concentration, one might be translated by the ribosome at a greater rate than the others – in other words, that some control is exerted during the functioning of the ribosome.

So we have three potential points of control: the rate of synthesis of the messenger RNA corresponding to a particular protein, the rate of degradation of this messenger RNA, and the rate of translation of the messenger RNA by the ribosome. There is much more experimental evidence that relates to the first of these possibilities than to the other two, and in the bacterial system that we shall describe it is the control of messenger RNA synthesis that is more or less exclusively involved in regulating the synthesis of specific proteins. However, in higher cells the other two modes of control may well operate to an important extent (see below). We must emphasize, then, that the one system that we shall describe is not to be taken as a model for all regulation of protein synthesis. The reason that we discuss it is because it exemplifies one possible means of controlling protein synthesis in bacteria, and because it is the result of an exceptionally fruitful and brilliant collaboration between biochemistry and genetics.

When wild-type cells of *E. coli* are grown in glucose medium, they make extremely small quantities of β-galactosidase (the enzyme that catalyses the hydrolysis of lactose to glucose and galactose). When lactose replaces the glucose, the rate of synthesis of β-galactosidase increases by many hundred times. This phenomenon (the increase in rate of synthesis of an enzyme) is called induction, and a substance that, when present in the medium, causes induction is called an inducer. Although lactose is a powerful inducer of β-galactosidase it is not the most powerful known; various synthetic compounds, all of which are chemical analogues of lactose, are inducers.

β-Galactosidase is essential to enable cells to grow on lactose, and if one selects mutants of *E. coli* that cannot grow on lactose one finds that most of them are unable to synthesize β-galactosidase. Some of them, however, are lacking not β-galactosidase, but another protein which is called galactoside permease and which is essential for the uptake of lactose from the medium into the cells. Experiments with inducers have shown that β-galactosidase and galactoside permease are, in wild-type cells, always induced *together*. In the absence of inducer both are made at a very low rate; in the presence of inducer both are normally made at a very high rate; but in the presence of a low concentration of inducer, sufficient to cause only partial induction, both are made at an intermediate rate. The synthesis of the two proteins must therefore be controlled together.

It is possible to find another kind of mutant that is able to synthesize both β-galactosidase and the permease but is unable to *control* this synthesis. As a result both proteins are made by these mutant cells even in the absence of inducer; this phenomenon is called constitutive synthesis and the strains are called constitutive mutants.

Now by genetic techniques it is possible to map the genes for β-galactosidase and for galactoside permease, and these are found to be contiguous. It is possible, too, to study genetically the mutants that synthesize the two proteins constitutively; each of these is found to have a mutation in a gene that is located near to, but not contiguous with, the genes for β-galactosidase and for galactoside permease. Moreover, it is possible to prove (also by genetic techniques) that wild-type strains are capable of synthesizing a substance which constitutive strains cannot synthesize, in other words that a strain is constitutive *for want of* this product. It follows that in wild-type strains in the absence of inducer the synthesis of β-galactosidase and galactoside permease is normally prevented or *repressed*; the sustance that acts in this way is known as the repressor. The repressor is found to be a protein, which is absent from constitutive strains.

So far we have described three proteins, β-galactosidase, galactoside permease and the repressor, and their corresponding genetic elements. In addition the system contains two further genetic elements which are believed not to specify any protein products. These are the promoter and the operator, the functions of which we shall discuss below. The five elements are arranged along the DNA in the order:

repressor gene – promoter – operator – β-galactosidase gene –
galactoside permease gene

and we can now give a complete description of the working of the system.

In the absence of a galactoside inducer the repressor (which is the protein product of the repressor gene) binds to the operator DNA. This highly specific interaction between protein and DNA prevents transcription by the RNA polymerase, which can initiate synthesis of messenger RNA only by binding to the promoter site on the DNA. Hence in the absence of an inducer no specific messenger RNA is formed. Inducers act by becoming attached to the repressor. This complex of inducer and repressor can no longer bind to the operator; in these circumstances the RNA polymerase can bind to the promoter and synthesize a messenger RNA molecule that corresponds to both β-galactosidase and the permease. This messenger RNA is translated by the ribosomes, and thus in the presence of an inducer β-galactosidase and galactoside permease are synthesized.

We remarked earlier than this system must not be taken as a model of all control systems, even in bacteria. However it contains several features that are highly characteristic of protein synthesis in bacteria. You will see that the messenger RNA that is synthesized from the promoter corresponds to *both* β-galactosidase *and* galactoside permease. A messenger RNA of this sort, corresponding to more than one protein, is called a polygenic messenger RNA. Such polygenic messenger RNA molecules are common in bacteria: a striking example is that for the histidine-biosynthetic enzymes, where the messenger RNA corresponds to ten enzymes. Plainly in systems involving either biosynthetic or degradative sequences it is extremely desirable for the synthesis of all the enzymes to be controlled together: either histidine is present in the medium, in which case none of the biosynthetic enzymes are needed, or it is absent, in which case all of them are needed together. It is presumably for this reason that the genes for a particular pathway of synthesis or degradation often lie together (p. 153).

Another general point that we may derive from considering the 'lactose' system is that each gene, or set of genes that are transcribed together, probably has its own promoter site at which the RNA polymerase binds and begins transcription. If these sites vary in affinity for the polymerase, one can readily see how it is that different genes or sets of genes are expressed at different rates. For example, the repressor gene is expressed at a very low rate, since repressor needs to be made only in extremely small quantities (there are only about a dozen repressor molecules in each cell, as compared with about 10^5 molecules of β-galactosidase in a fully induced cell); this suggests that the repressor gene has a poor promoter.

There is a further feature of the 'lactose' system which we have not hitherto mentioned but which seems to be of general applicability in bacteria. When a galactoside inducer is added to a culture of bacteria the synthesis of β-galactosidase and the permease begins very quickly; equally when the inducer is removed synthesis ceases very quickly. The rapid cessation seems to be due to rapid breakdown of messenger RNA, and in general it is found that in bacteria most messenger RNA is degraded very rapidly. This is in part the reason why bacteria are able to adapt quickly to changes in the medium by changing their pattern of protein synthesis: once a particular species of messenger RNA is degraded the ribosomes that were translating it will be free to translate newly synthesized messenger RNA (see p. 167). (Thus in order to maintain induction of a particular set of genes, the corresponding messenger RNA must be synthesized continuously.) In higher cells, however, there is no corresponding need to adapt to changes in the medium (see below), and messenger RNA appears not to be degraded rapidly. It may be that control of protein synthesis is exerted in part through different rates of degradation of different species of messenger RNA, but there is no firm information about this possibility.

We suggested above (p. 177) that aside from changes in the rate of synthesis and degradation of messenger RNA, protein synthesis might be regulated in part at the level of translation of the messenger RNA by the ribosome. One interesting example is that of polygenic messenger RNA. If there were no control of the extent of translation, the protein products of a polygenic messenger RNA would be synthesized in equimolar quantities; in practice it is found that this rule does not always hold, so that it seems that there is control of translation of the messenger RNA as well as of its synthesis (and conceivably its degradation). There is reason to believe that in higher cells the control of translation plays an important part in regulating protein synthesis as a whole.

In fact the control of protein synthesis in higher cells is a rather different phenomenon from that in bacteria. We suggested at the beginning of this chapter that it is advantageous for bacteria to be able to change quite abruptly the rate of synthesis of some enzymes in response to changes in the environment. In multicellular organisms abrupt changes in the environment of the cells do not normally occur (see also p. 184), and the same kind of rapid change of pattern in protein synthesis is not needed (although induction of enzymes does happen to a modest extent in animals). What is important for these organisms, by contrast, is to be able to *differentiate* their cells, inasmuch as different tissues specialize in the synthesis of different proteins. Despite this differentiation almost all cells in the organism retain a complete complement of genetic material; a cell in the pancreas, for example, has the genetic information necessary to make haemoglobin although this is never expressed. The need, then, is for cells to be able *permanently* to repress the expression of much of their DNA. The bacterial control systems that we have described may well prove to be only a very distant analogy for this kind of process, if an analogy at all.

Control of enzymic activity

In the chapters on intermediary metabolism, we described pathways for the degradation and synthesis of several compounds. We outlined, for example, the route of degradation of glycogen, and also the route of synthesis (which, as we remarked, is slightly different). But any given cell, even though it contains the enzymes of both pathways, will not be rapidly degrading and synthesizing glycogen at the same moment. There must therefore be some control system that enables one pathway to be active, and the other inhibited, according to requirements.

In a similar way there must be means of preventing the synthesis of intermediates or building-blocks that are not required. To refer again to histidine synthesis in bacteria, we have already seen that the addition of histidine to the medium represses the formation of the enzymes required for histidine synthesis. However, the enzymes that are already present in the cell will be expected to be active for a comparatively long time, and will, unless there is a rapid means of rendering them inactive, waste both raw materials and energy in making a superfluous compound. In fact it is found that if an amino acid, purine or pyrimidine is added to a culture of bacteria it will often inhibit its own synthesis. This inhibition occurs so rapidly that repression of enzyme synthesis is insufficient to account for the observed result.

These examples suggest that there are means of inhibiting or enhancing enzymic activity, in addition to the means of repressing and inducing enzyme synthesis that we described in the last section. The best-studied examples are those that we mentioned in the preceding paragraph, namely the inhibition by amino acids, purines or pyrimidines of their own synthesis (often called end-product inhibition), and we shall say something about this first. Later we shall apply some of these principles in describing the way in which the degradation and synthesis of carbohydrates is controlled.

Two important characteristics of end-product inhibition are that the end-product inhibits not all of the enzymes leading to its synthesis but only the first enzyme that is specific to the pathway, and that this inhibition is a property of the affected enzyme itself and can be observed outside the cell in a purified enzyme preparation. In both these respects end-product inhibition differs from repression. A detailed discussion of end-product inhibition would require an account of enzyme kinetics, which is beyond the scope of this book; but we can outline some general principles without considering in depth the behaviour of the inhibited enzyme.

One of the best characterized of the enzymes that are subject to end-product inhibition is aspartate transcarbamylase, which is the first enzyme in the pathway of pyrimidine synthesis (see p. 157). Notice that the $\Delta G^{0'}$ of the reaction catalysed by this enzyme overwhelmingly favours the synthesis of carbamyl aspartic acid. This is an important feature of reactions that are subject to end-product inhibition. We have seen throughout this book that 'reversible' reactions (that is, those in which $\Delta G^{0'}$ is small enough to allow the formation of a significant amount of products in either direction) are generally used both in degradative and in synthetic pathways. Therefore to subject an enzyme that catalyses one of these reactions to end-product inhibition, or indeed to any other inhibition or activation, would have no effect in controlling the rate of degradation or synthesis as a whole. We shall refer to this point again in discussing the control of degradation and synthesis of carbohydrates.

The enzyme aspartate transcarbamylase is inhibited by CTP, which is one of the end-products of the pathway of pyrimidine synthesis. The effect of CTP on the enzyme is an interesting one. The first stage in an enzyme reaction is binding of the substrate (see p. 45); and this enzyme has to bind both aspartic acid and carbamyl phosphate. Now it is found that in the presence of CTP the enzyme undergoes a change in its quaternary structure, and the result of this is that its affinity for aspartic acid is much diminished. Thus at a given concentration of aspartic acid the rate of reaction is much less in the presence of CTP than in its absence. (The change in the configuration of aspartate trans-carbamylase is reversible, so that removal of CTP will restore its activity to normal.) The molecular basis for this phenomenon is analogous to the mechanism by which the binding of one molecule of oxygen to haemoglobin alters its affinity for subsequent oxygen molecules (see p. 39).

This kind of enzymic property is obviously useful in an organism that may suffer changes in its environment. For example a bacterium will, in the absence of pyrimidines from the medium, make its own pyrimidines by the pathway that we outlined on pp. 157-9, which includes the reaction catalysed by aspartate transcarb-amylase; if a pyrimidine is supplied in the medium some of it will be converted to CTP and this will inhibit the enzyme. But in addition end-product inhibition is an extremely valuable control for *internal* regulation of a cell's metabolism. Suppose, for example, that the rate of RNA synthesis in any cell is, for some reason, suddenly reduced. This reduction will result in an accumulation of the precursors of RNA synthesis, one of which is CTP. End-product inhibition will now diminish the rate of pyrimidine synthesis until the concentration of CTP falls to its normal level, when pyrimidine synthesis will be quickly resumed.

Changes in affinity of enzymes for their substrates are a common means of regulating metabolic activity. They are frequently brought about by end-products (as with aspartate transcarbamylase) and then are generally inhibitory – i.e. the end-product diminishes the affinity of an enzyme for its substrate. Sometimes, however, they are brought about by other substances, and they may then involve either a decrease or an increase in affinity. Interesting examples come from studies of the control of glycolysis and of carbohydrate synthesis.

Glycolysis is a process that requires control for reasons of the cell's internal economy. There is no end-product which, like a pyrimidine, may suddenly turn out to be in large excess in the medium that bathes the cell. On the other hand glycolysis is crucial for supplying energy and thus in regulating, in a very general way, all the cell's activities. Its control is obviously essential in metabolism.

Now although glycolysis has no true end-product, there is a sense in which ATP can be regarded as its end-product. In a situation where ATP is abundant the need for glycolytic activity is obviously greatly reduced. Conversely when ATP is in short supply, glycolysis requires to be stimulated.

If we compare the glycolytic pathway (Ch. 10) with the pathway for synthesizing carbohydrate (Ch. 15), we shall see that most of the reactions occur in both – that is to say that they can achieve a net formation of products in either direction. There is, however, one reaction, placed at a crucial stage, that has a high negative $\Delta G^{0'}$, namely the phosphorylation of fructose-6-phosphate by ATP, catalysed by phosphofructokinase. Whether the substrate of glycolysis is glucose or glycogen or galactose, phosphofructokinase provides a potential point of control for the whole process.

It is therefore not surprising that phosphofructokinase is inhibited by ATP. Moreover, the adenine nucleotides in the cell are in equilibrium owing to an enzyme-catalysed reaction:

$$ATP + AMP \rightleftharpoons 2ADP$$

so that when the concentration of ATP is high the concentration of AMP is low. Phosphofructokinase is found to be greatly stimulated by AMP as well as inhibited by ATP (which is also one of its substrates).

Exactly the converse arguments apply to the synthesis of glycogen from lactic acid or from intermediates in the Krebs cycle. Under conditions where the supply of ATP is plentiful, and therefore (see the reaction above) the concentration of AMP is low, it is desirable for the cell to be able to synthesize carbohydrate from simpler precursors and thus build up its stores of substrate. When the concentration of ATP falls, and that of AMP rises, it is desirable to inhibit the synthesis of carbohydrate. Now we have seen (p. 127) that one of the reactions that are specific to carbo-hydrate synthesis is the hydrolysis of fructose-1;6-diphosphate to fructose-6-phosphate catalysed by fructose diphosphatase; and it is found that this enzyme is strongly inhibited by AMP.

Our final illustration of the way in which enzyme activity is controlled is the breakdown of glycogen. The reaction that degrades glycogen to glucose-1-phosphate is catalysed by the enzyme phosphorylase (p. 95). Phosphorylase in animal muscles exists in two forms, *a* and *b*, of which the *a* form is the more active. Phosphorylase *b* can be converted into the *a* form by a rather complicated series of reactions, and the effect of this conversion is to increase the total activity of the enzyme and thus to ensure that glycogen is more quickly broken down. Now, one of the agents that activates the system which converts phosphorylase *b* to phosphorylase *a* is the hormone adrenalin. This, then, is an example in which we can define the action of a hormone (the *physiological* effects of which have been known for a long time) in biochemical terms.

Throughout this book we have been at pains to stress the fact that, while many biochemical reactions can achieve a net formation of products in either direction, there are several that cannot (without a source of energy) because their negative $\Delta G^{0\prime}$ is extremely high. The examples that we have given in the last few pages illustrate how these latter reactions can be controlled, and how changes in the activity of the enzymes that catalyse these reactions can control whole pathways. By contrast reactions such as the isomerization of the triose phosphates generally belong both to a degradative pathway (p. 91) and to a synthetic pathway (p. 126), and to control this reaction would have no useful effect. These principles are at work in control of the synthesis and degradation not only of the major substrates (carbohydrate and fat) but also of the building blocks of macromolecules (amino acids, purines and pyrimidines) and of nucleic acids and proteins.

Index

acetaldehyde, 94
acetoacetic acid, 107
acetoacetyl coenzyme A, 107, 142
acetone, 107
acetyl coenzyme A, 71, 97, 111
  carboxylation of, in synthesis of
    fatty acids, 131–2, 133
  from fats, 105, 106, 130
  from isoleucine, 142, 143
  in Krebs cycle, 98, 99, 101
  not converted to Krebs cycle
    intermediates, 130, 142
acetyl coenzyme A carboxylase, 131
acetyl fragments, in intermediary
  metabolism, 84, 85, 86, 87
acids (two- to six-carbon), in intermediary metabolism, 84, 85, 86, 87
aconitase, 98
actin, muscle protein, 32
acyl carrier protein (ACP), 132, 133,
  134
acyl mercaptide (high-energy) bonds,
  71, 99
acyl phosphates, high-energy compounds, 71, 91
adenine, 51, 154
adenine nucleotides, equilibrium between, 185
adenosine diphosphate (ADP), 69, 70,
  74, 81, 131, 185
  synthesis of, 156–7
adenosine triphosphate (ATP), 69–71,
  74, 131
  inhibits phosphofructokinase, 184–5
  in muscle, 32
  produced: in fat breakdown, 103,
    105–6; in glycolysis, 92, 93,
    94; in intermediary metabolism,
    83, 84; in Krebs cycle, 99,
    100–1; in oxidation of glucose,
    100–1, 106; in oxidative phos-
    phorylation, 73, 80, 81; in photo-
    synthesis, 69, 85, 116–18.
  required: in carboxylation of py-

ruvic acid, 101; in fixation of
ammonia, 155; in phosphorylation of fructose-6-phosphate, 90,
and of nucleoside mono- and diphosphates, 129, 157, 159; in
photosynthesis, 115, 120, 122; in
reaction of fatty acids with coenzyme A, 103, 106; in synthesis
of amino acyl tRNAs, 144, 174, of
fats, 131, 134, of polysaccharides,
130, of proteins, 144, and of
purines, 156, 157
adrenalin, 185
alanine, 16, 83
  interconversion of pyruvic acid and,
    124, 138–9, 143
aldol condensation, 121
aldolase, 91, 121, 126
amino acids
  activation of, 144–5
  essential, 27, 140–1, 150
  metabolism of, 136–43
  in proteins, 5, 7, 8, 15–19
  tRNAs for, 54, 167–9
  synthesis of, 86, 87
  synthesis of proteins from, 165–74
amino terminus of protein, 5, 172
aminoacyl-AMPs, high-energy compounds, 144, 167
aminoacyl transfer RNAs, 54, 167–9,
  174
ammonia
  from amino acids, 138, 141, 143
  fixation of, 136–8, 155
  from urea, 42
amylopectin, 56
amylose, 56, 129
anaemias, 40
antibodies, 34–6
anticodons, 170
antigens, 34–6
arginine, 17, 19, 21, 42
  enzymes for synthesis of, 153
  in intermediary metabolism, 140, 143